DECYCLE

A NEW WAY TO STOP MILLIONS OF TONNES OF EMISSIONS

ROBERT SETTER

Decycle
Copyright © 2023 by Robert Setter

Tellwell Talent
www.tellwell.ca

ISBN
978-0-2288-9444-5 (Paperback)
978-0-2288-9445-2 (eBook)

TABLE OF CONTENTS

INTRODUCTION

The time is right, if not overdue, for innovative, practical solutions to directly address global warming. I have watched hundreds of news stories, investigative reports, and read dozens of articles on the topic, and am consistently struck that so many end the same way. After articulating the case that concrete action is needed to avert flooding, fires, and droughts, including the financial costs and impact on humans, invariably, at the conclusion, the plea is for novel solutions to help the planet.

On April 4, 2022, the Intergovernmental Panel on Climate Change released a report based on roughly 18,000 studies, approved by 195 countries concluding carbon emissions need to peak by 2025 and fall by 43% by 2030 to limit global warming to 1.5 degrees Celsius. Point five of the report for action notes a reduction in material consumption is one of the central tenets.

This book covers an implementable practical solution that requires little capital or decades of study to achieve two goals: permanently stopping carbon emissions and saving money as city collection expenditures are reduced. In summary, Decycling intercepts the circuitous route recycled material currently takes. Consumers and cities can stop chauffeuring their waste material around town and instead leave excess packaging at the point of purchase, where it can be processed within existing transportation infrastructure. Decycling includes the participation of consumers in waste diversion that will permanently stop greenhouse gas emissions. The broader the scope of adoption around the globe, the greater the amount of carbon emissions stopped.

The adoption of Decycling by Europe and North America alone would save 2.5 million tonnes of carbon emissions per year.

At this point in our economic evolution, it makes sense to identify embedded inefficiencies in current systems where relatively minor adjustments can yield significant measurable results in reducing pollution and carbon emissions. The Decycling difference compared with other initiatives is cost. Carbon capture schemes, solar, wind, and electric transport are expensive. All of the aforementioned initiatives have merit, and are being utilized, however, costs tend to be prohibitive which inhibits robust uptake. Most produce waste at the end of their life cycle which creates yet another environmental challenge. It is time we stopped the outdated practice of chauffeuring waste material all over town.

One aspect of this book is to reframe where we are now and leave the reader more hopeful about the future. It is well documented that young people, specifically, are increasingly moored in a sense of anguish and anxiety given the constant barrage of news that global warming could threaten their way of life and life on earth in the coming decades. The points covered in section one aim to help ease the anxiety and create at least a modicum of hopefulness; it is not too late, much can still be accomplished.

Section two is a reminder of where we are in the grand scheme of the cosmos and includes a look back at practices that were actually very "green." Section three reviews many of the current initiatives; all help and have value but are costly and come with hidden emissions along with end of life disposal issues. Nothing is flawless; however, there are many current technologies and new ones coming that will help deliver the needed changes. Section four details what it means to Decycle including an estimate of the carbon emission savings possible.

There is no reason to think we can't get there; humans have been remarkably effective in finding solutions and acting when the stakes are understood. Decycling intersects embedded inefficiencies as a practical innovation that does not require large capital investments. Most importantly, Decycling involves the participation of all of us — the consumers.

The main affliction of our modern civilization is that we don't know how to handle the suffering inside of us and we try to cover it up with all kinds of consumption. Thich Nhat Hanh

1

A LOOK AHEAD

IT'S UP TO US

It all circles back to the actions of the individual in taking on personal responsibility. Too often, we tend to think, oh well, what does it hurt if I do this even though I know I probably shouldn't? It's just me, here, alone, one time; nobody will know, so how could that really matter in the grand scheme of things? It's as if we are born with this ability to rationalize behaviours when convenient; then again, maybe it is just learned from observation as we grow up into teens and then adulthood.

I can recall hearing this assertion from my kids, "you should see my friend's house and his room; it is way worse than this, and you say this is a mess and want me to clean it up now?!" Moral relativism may be at the core of why we are where we are now. The challenge is perhaps best illustrated by this classic example of human behaviour I first heard during a lecture given by one of my Economics professors in 1987.

It is the conundrum of "walk across the grass or don't walk across the grass" as follows. A woman moves into a house on a corner lot and notices the boulevard at the corner of her property is bare, so she decides to spruce it up by planting grass. She then notices pedestrians taking a shortcut through the just seeded area, so she fashions a string fence around it with a sign indicating it is newly seeded; please stay off.

When the pedestrians approach this new string barrier, they pause for a moment to consider the request but then think, "Oh well, whatever, just me crossing through one time won't matter." Human nature kicks in to rationalize that disobeying this tiny request one time will not hurt anything. Implicit in the individual's thinking at that moment is that everyone else will respect the sign and take four extra steps to go around. A few weeks later, the grass comes in over the entire area except for a bare pathway through the middle of the grass perfectly bisecting the corner space. The bitter truth is we cannot rely on everyone else to do the right thing so we can get a free pass.

ANYTHING IS POSSIBLE

Proof for me that optimism towards the future is grounded in reality is the once catastrophic issue of the vanishing ozone layer, which dominated the news cycle a few decades ago, much like global warming does today. It was a colossal problem; the future of the planet was threatened. This ever-pressing issue became embedded in our collective psyche and even earned a mention in a rather intense Neil Young song from that period. Today, we see that what was considered a planet-ending catastrophe has now been vanquished, according to several news reports and academic papers. In summary, the ozone layer appears to be closing at both poles.

How did this change in fortune occur? Scientists determined it was emissions from refrigeration units emitting CFC gases which then dispersed into the atmosphere and reduced the density of the ozone layer. It is an essential part of the atmosphere, integral in shielding planet earth from powerful UV rays. The solution was simply to stop using those gases which meant new systems for cooling were needed. Evidently, efforts proved successful. Further, some governments incentivized owners of older refrigeration units to retire them by offering free pickup along with a monetary incentive. We can breathe a collective sigh of relief today because quick and concrete actions were taken back then. The ozone layer over the Antarctic had been diminishing since the 1970s, but news the hole now covered the entire Antarctic continent triggered worldwide alarm. In 1987, world leaders signed the landmark Montreal Protocol, hailed as one of the most successful environmental treaties of all time. Ozone-depleting chemicals were phased out, with industry switching to "CFC-free" aerosol cans. "It was a global problem, but industry, the scientists, the policymakers came together," says Dr. Jones.[1]

That's not the only one, there have been other environmental challenges that were acted upon with impressive results and success. Sulphur dioxide fallout from coal burning plants was termed "acid rain" for the damage it was causing to the environment. "'In the '80s, essentially the message was that this was the largest environmental problem of all time,' says Peringe Grennfelt, a Swedish scientist who played a key role in highlighting the dangers of acid rain. Headlines warning of the threats of acid rain were commonplace. For years there had been obfuscation, denial and diplomatic stand-offs, but once the science was settled beyond doubt, calls for action quickly gathered momentum. It led to international agreements curbing the pollutants from burning fossil fuels that acidify rain."[2]

Flash forward to today, where we have a new ominous threat that many think won't be solved in time to avert irreversible disaster. On the positive ozone news, there doesn't seem to be the amount of mainstream coverage one might expect to see. It seems the mainstream media, (MSM), prefers to promulgate doom-and-gloom narratives to keep viewers in a perpetual state of anxiety. It's possible I missed it, or maybe there was no room for it between the other stories on planetary demise, inflation, war, crime, health catastrophes, and political craziness.

HOPE IN TECHNOLOGY

Over time, it is probable that technology will deliver us from the issues plaguing the planet as we forge ahead into decade three of the new century. As recent evidence, technology just delivered the planet from a pandemic with the speedy development of new treatments and vaccinations, although it is now known that the vaccine efficacy was over-promised according to recent data. Still, the primary goal of preventing catastrophic personal outcomes was met and serves as some proof that when humans are pressed with their backs against the wall and agree to work together, share resources, and unite in a common goal, great feats can be achieved in a timely manner.

An innovative technology I recently came across online confirms the ability to extract hydrogen and oxygen from the atmosphere and condense it into drinking water. The unit that accomplishes this is called a Phantor and is powered entirely via solar. The water passes through carbon bacterial filters and is disinfected with ultraviolet light. The unit can produce up to ten thousand litres daily and be transported to regions where drinking water is suddenly cut off. Knowing this type of technology already exists, one can envision a future of scaling up to address the lack of water which is currently an issue for some two billion people on the planet.[3]

When examining transportation, the ideal situation is no gas, no charging, just self-generating power. A vehicle like this does exist today. Engineers have already created solar-powered vehicle prototypes, which is very promising, especially for those living year-round in sunny climates. As they are driven, the sun continuously provides the energy needed for propulsion while the battery stores the energy required for driving at night or on darker cloudier days. Manufacturers are moving quickly with at least one planning a 2024 EV that will have a solar panel on the exterior of the body. It is encouraging to see practical real world usage now and portends where transportation options are heading.

SOLAR FILM

Innovative solar applications are another interesting and positive development. A few years back, while working as a contributing writer to *Green Life Living* magazine, I interviewed the founder of a company named Q-Solar. The company had developed and was in the process of testing a thin film that functioned as a solar panel. Envision peel-and-stick decorative items for windows, except in this case, the entire window would be covered, so as the sun hit the window, the solar energy would be captured and used to power the building. The application was also intended to cool the building as needed.

Q-Solar is no longer around. It is unclear if it was acquired or ran out of funds; however, I recently heard a brief radio interview with a California-based enterprise developing a product that sounded similar in function. Q-Solar likely ran into the fundamental cost benefit analysis issue; the price to manufacture and install the product was greater than the monetary return realized from the reduction in energy usage. As an early-stage company, it was likely not financially viable to continue.

However, work on this type of solar application has continued with a transparent solar panel design currently in development at the Copenhagen International School. Their goal is to place panels over the entire surface of the building in replacement of standard glass windows, not just cover them. It has been reported that these clear window-like panels can produce over half the energy the building requires.[4]

Similarly, developments continue with specially designed photosensitized dye molecules capable of harvesting photons from across the entire visible light spectrum. The idea is to use them in buildings to replace the traditional glass currently in use. This high-tech replacement glass is reported to be affordable to produce and is pliable. The first commercial installation has been undertaken at the SwissTech Convention Centre, but thus far, the ability to generate electricity is not yet substantive and is not as efficient as today's solar panels. Regardless, this remains another promising ongoing development that will only improve with continued research.

SOLAR EVERYWHERE

Alternative solar applications, such as solar roads and sidewalks, are now moving past the testing phase. These are streets and sidewalks made with solar panels in place of cement or asphalt. Solar roadways have shown some efficacy but cannot match the capability of solar farms as roads quickly become covered in dust and debris, plus the traffic moving along the roadway covers the panels preventing the capture of UV rays.

A recent update on the reconstruction of a town in British Columbia that fell victim to catastrophic forest fires during the summer of 2021, noted the rebuild would include solar sidewalks to capture passive solar energy as part of a plan to rebuild the town to net zero emissions status. "Solar Earth Technologies makes solar panels that not only generate electricity to power homes but can withstand the weight of a five-ton vehicle… The hardened solar panels can be fitted into sidewalks, trails, bike paths, driveways, and parking lots."[5] As of June 2023, the rebuilding process is moving forward but not as quickly as initially anticipated.

MICRO HYDRO

Where I reside is typically wet from October to March, and recently, summers have seen long dry stretches with zero rain for many weeks. One January night a few years ago was much rainier than usual as a weather system termed a "pineapple express" produced an ongoing torrential downpour. The weather systems' fruity name comes from the fact that these heavily moisture-laden lows originate near Hawaii before sweeping North, where copious amounts of rain and warm air are delivered to the Pacific Northwest.

On that January night, I was standing outside under cover to experience the warmth, but after a minute or two, it was the sound that caught my attention; not the pounding rain but the gushing sound emanating from the drainpipes as the water cascaded down from the roof and eaves troughs. I then wondered what if the energy from the rushing water could be captured and if so would it even be worthwhile?

Initial research quickly led to the fact that an American University had created a prototype to determine if micro hydro could be a viable way to provide a small amount of energy for use in rural Africa.[6] As technology has spread across the African continent some rural farmers with near subsistence living standards have increased their incomes by using a cell phone connected to the internet. Once connected to the online world, they had access to banking, and with banking, commercial enterprise. However, with sparse availability of electricity sources in rural areas, the practicality of setting up e-commerce for rural African communities faced roadblocks.

When the researchers took the device into the field for testing, they were pleased to see that it did indeed work to produce enough electricity to charge a cell phone. The world now has proof of concept with a prototype that uses rainwater runoff funneled through tiny turbines to create electricity. For locations with considerable amounts of annual rainfall, it is perhaps foreseeable that at some point, new homes will have downspouts fitted with a series of micro hydro generators to capture even more of the free power that the earth's climate provides.

The REDAWN project (Reducing Energy Dependency in Atlantic Area Water Networks) is looking at the feasibility of hydropower energy recovery technology by building it into existing water distribution systems. Currently, there are fifteen partners from five different countries working on it. This differs from micro hydro as the installations are aimed at municipal water systems where much greater power generation is possible than with home based micro hydro. As developments continue, it does become easier to envision what clean energy creation will look like in the decades ahead.

A big plus with this type of energy capture is the low environmental impact compared to other existing "green" initiatives. Although building hydropower recovery technology into existing water distribution systems is still uncommon, it is yet another innovative technological development that will be a new source of "green" energy. Capturing all the free energy available is a direction the world must continue to move in provided the cost benefit return is sound.

MAGNETIC PODS

Another intriguing technological development, named the Hyperloop train, uses magnetically levitated pods to transport people in a vacuum tube at speeds faster than air travel. This technology is being advanced in many countries including China. Ground travel at speeds faster than air does sound futuristic and as though it would be decades away; however, phase one construction for a metro line of this type is scheduled for 2023 in the city of Edmonton. The line will transport people from downtown to the airport, with additional plans to build a line between Edmonton and Calgary. The proponents state that it will use less energy than an electric vehicle in getting people from point A to point B and will be faster and cheaper than air travel. The main drawbacks are the initial costs to build and the ongoing issue of ensuring the vacuum tube is kept at a consistently low pressure, which does require a substantial amount of energy. The reasoning is that once the system is up and running, it will replace flights between Calgary and Edmonton which should reduce carbon emissions overall.

CLOUD SEEDING

Drought continues to be in the news although the winter of 2022-23 saw drought relief in the Western US as a steady stream of storms hit the west coast. If not here in North America, the storyline shifts to Africa, Asia, Australia, or some other location void of rainfall for months and sometimes years. The narratives tend to be similar, noting that the arid conditions are related to global warming and there is little humans can do about it except reduce carbon emissions in hopes of combatting these prolonged dry spells. For some time now, scientists and engineers have been working on ways to force clouds to emit rain and snow. Measurements have shown that an increase in precipitation of up to 15% is achievable. In the past few years, a small but growing private-sector cloud-seeding industry has emerged and is advancing this solution.

The process of cloud seeding involves injecting particles into clouds with the goal of triggering or releasing rainfall. The process has a track record of efficacy; though, not in a consistently predictable way and would appear to have decades of development ahead before it will be anything more than a helpful tool for drought and water management. China is the global leader in developing this technology, although cloud seeding is also taking place in America, Australia, and other arid regions around the globe. In fact, in the summer of 2023 Mexico used cloud seeding to combat drought and noted it was a net positive. It was estimated that the treated clouds produced 30 percent more rain however some observers remain unconvinced.

One positive from an environmental standpoint is that studies have so far shown that fallout from adding seeds to clouds has no documented negative environmental impact. "The technology is steadily improving. Most notably, scientists now have computer models to simulate the effects of silver iodide seeding, allowing scientists to pinpoint the most suitable locations for weather enhancement. As processing power improves, so will these models. But no matter how sophisticated the modeling becomes, the method has little chance of helping during a drought emergency — mainly because you need storm clouds for successful seeding, and droughts by nature offer few of them."[7]

GEOTHERMAL

One of the most positive characteristics of geothermal is the current belief that this energy source is by all accounts renewable, as it harvests energy emanating from the core of the earth. The notion of sustainability is considered sound because the amount of energy extracted is minuscule compared to the total amount believed to exist beneath the planet's surface. Therefore, though it is not renewable in the basic definition of the word, because the amount of energy harvested is considered inconsequential when compared to the total stock available, it has been granted this label.

Although geothermal has yet to be widely implemented, there are some installations in a number of countries, with exploration and development continuing in an effort to determine effectiveness on larger scales. In the 1980s, the Canadian government committed resources to the exploration of viability. However, a few years later, it lowered the priority and placed it on hold although the private sector recently revived geothermal in Canada. A company based out of Calgary is now working on developing geothermal plants and is seeking to access dormant volcanoes situated north of Vancouver, BC. The location is considered ideal for geothermal due to its proximity to the highly populated greater Vancouver metropolitan area.

In the meantime, another company operating in Saskatchewan is moving ahead with the construction of a geothermal plant to supply energy for up to 25,000 homes. Phase one is expected to produce five megawatts of energy when completed in the summer of 2024. Geothermal is more expensive to build than alternatives like wind turbines and solar but has the advantage of being a steady source of power around the clock.

Some estimates indicate geothermal could supply up to five percent of global energy within 25 years and possibly up to ten percent by the end of the century. The carbon emissions footprint is estimated to be 95% lower than coal-fired plants and therefore shows tremendous promise as an alternative energy, although it does have a long way to go before its full potential can be realized. It is entirely possible that by 2050, we will be tapping the inside of the earth to save the outside of the earth in a greater magnitude than is currently forecast.

WATERFRONT MORTGAGES ARE AVAILABLE

According to NASA's estimates, ocean levels have risen about ten centimetres since 1993. They note that the rise is caused by two main factors; melting ice from glaciers and warming seawater, which expands as it warms. Two systems of measurement are cited; one is from satellite data, and the other is from coastline tide measurements that date back to the beginning of the last century. From 2015 to 2020, the increase has been measured to be about two centimetres or just under one inch. If that rate of change remains constant, by 2050, sea levels will be 12 cm higher than today, which is approaching half a foot at just under five inches.[8] (4 cm per decade x 3 decades)

One would think that rising sea levels will be a major issue for houses in low-lying areas near the ocean and especially for waterfront properties. Yet, real estate advertisements continue to promote the dream of waterfront living. It is curious that even now, waterfront properties do not sell at a discount but instead still command a premium. The topic of rising sea levels piqued my curiosity roughly twenty-five years ago when I first moved to the east coast of Vancouver Island. When the inland highway was completed, the vast majority of traffic using the coast highway rerouted as expected. As vehicle traffic fell off, business activity cratered, resulting in extremely attractive prices on waterfront properties along that stretch of highway. I recall thinking that maybe these should be marketed as soon-to-be underwater properties, but that doesn't have much of a ring to it. Then again, apparently nothing noticeable has happened either, at least here.

This is either another reason to be hopeful about the future or is a flagrant financial system oversight. It could be the internal research of financial institutions indicates there will not be a significant impact from rising sea levels until after the multi-decade mortgages being issued have been paid back, at which time it is no longer their problem. Then again, insurance is another matter now in the news as it has been reported that in some southern coastal states, property insurance companies are beginning to back away citing rising water levels and stronger more frequent storms as the reason.

Researchers have also noted that while some coastal properties may not flood, rising sea levels could make an impact by cutting them off as lower lying areas nearby are flooded at high tide. In that case, driving in and out won't be possible and there could be other implications including utilities and emergency services. Time will tell.

Part of the misconception around rising water levels is the understandable perception that the large bodies of water on planet earth behave like we would imagine from our own experience watching a sink, bathtub or pool fill up. As the water enters, it rises up all sides equally. This is not the case with bodies of water on planet earth for several reasons, some of which include tides and currents. Catastrophic flooding can happen in some places while other locations may experience little or no impact.

2
REALITY CHECK

HUMANITY IS VULNERABLE EVERY DAY

Amidst the complex demands of modern life, many have forgotten or are overlooking the fact that we reside on a water-covered sphere flying through space at enormous speeds which leads to the perpetual question; where are we actually situated in the grand scheme of the cosmos? In the time it took to read that last sentence, it is estimated you travelled roughly 2,400 kilometres.[9] You didn't move from your seat nor did you feel any forces other than the gravity which prevents all of us from floating away into space. It is hard to imagine that we really moved that far. The earth is spinning and moving around the sun, plus the earth and all is whirling around the centre of the galaxy which is also moving in space. Added all up and it's a lot to fathom. Scientists do their best to give us at least some sense that they are figuring it all out; however, back on earth one thing is certain, pollution of any kind is hazardous to the environment. For decades there has been a growing awareness of the need to preserve the planet in every way possible.

The present day preoccupation focuses on maintaining all modern conveniences and quality of life while simultaneously liberating the planet from ecological destruction in the quest for a nonpolluting world. Carbon neutral is the term commonly referenced in the current front page struggle, with the midpoint of this century cited by numerous sources as the year by which this must be achieved. Some believe the failure to do so will have catastrophic consequences. Others think it may already be too late, while another line of thinking points out that humans are highly adaptable and have already survived periods of cooling and warming so the worry is most likely way overdone.

Back to the question of what do we really know? Many data sets can be manipulated to illustrate whatever outcome the data publishers want. In the current world of information overload, a deeper examination past the sound bites and headlines is increasingly rare. Busy and distracted people hear thirty-second blurbs or read a six-word headline, then store the piece of information away before moving on to the next distraction. Too many trust the source, and give the mainstream media an automatic pass, and deem the information accurate. In contrast, alternate info-media sources are diminished or ignored because they don't fit the narrative, even when they feature reputable, highly regarded, well-educated professionals.

Those diligently seeking the truth are often left with their heads spinning, eventually checking out by default before reverting back to consuming the headlines as they fly by their screen of choice. Most people today lack the energy and time to delve deeper to determine what is factual versus what is plausible versus what is distorted. One of the most irritating instances of this misdirection occurs when polling results are announced with great reverence, as if they were facts, only to be followed by the methodology behind them; that it was an online poll, which renders the results completely worthless. Polling that provides accurate insights must be scientific, which means random and of a sufficient sample size, carried out impartially. Fortunately, reputable polling companies exist, but even the best-executed polls will be incorrect about five percent of the time.

I am old enough to remember the split in scientific opinions back in the 1970s when some scientists suggested the earth was cooling which led to concerns that food production would decrease, and energy use would increase. It was then discovered that long cycles in the earth's orbit might be the culprit responsible for the coming and going of ice ages. Adding in the air pollution that some thought was causing cooling, and you had some scientists of the opinion that another ice age was coming. Around the same time, new research on the impact of CO_2, which had been known about for decades, showed that the warming observed from these gases would be more than sufficient to offset any cooling trends. In 1976, *Time* magazine published an article about this, "The World's Climate: Unpredictable," which highlighted the lack of consensus on where the planet was headed long-term.[10]

When we awaken to a bright summer day, we head outside, gaze up and welcome the beautiful deep blue sky and bright yellow sun. All very much something most everyone enjoys; the sun is out, it's warm, plants are growing, and we can do dozens of activities outside while completely forgetting that the sun is a star like those we see on a clear night when looking out into the cosmos. What astronomers and physicists do believe as based on research, is that these hot, light-emitting balls of gas are not stable and will, of course, eventually run out of fuel.

When activity from the sun periodically ramps up, it often gets reported in the news. We are warned about increased sunspot activity possibly generating solar flares that could disrupt power and communications if they reach earth. In early 2023 news headlines proclaimed, "part of the sun breaks free and forms a strange vortex, baffling scientists."[11] This highlights how truly helpless humans are in the face of whatever our closest star might do. A few days later, another headline appeared declaring that a piece of the sun did not actually break free. Maybe it did, or perhaps not, but clearly something a bit out of the ordinary occurred.

Is it not a particularly unsettling notion to consider that when we observe stars millions of light years away, we are looking into ancient history? In our time-space continuum, those stars are no longer where we perceive them to be, as the light hitting our eyes on earth left those stars billions of years ago! Strange.[12]

What about the earth, and can it actually protect itself? Two MIT scientists found a strong pattern indicating this may be the case.[13] Over lengthy periods the planet may possess a feedback cycle that works to keep temperatures range bound such that life on earth survives. This is positive news, at least at first glance, and interesting to consider in light of what is happening today. Humans may not be able to ruin the planet as is now thought to be the case; however, the researchers clearly noted the periods involved span hundreds of thousands of years, so we better keep working on solutions to reduce pollution.

NASA has discovered a few things about global warming through monitoring sunspot activity and are clear that although sunspot activity is declining, the earth is not heading into a mini-ice age. "Global warming is the long-term heating of earth's surface observed since the pre-industrial period (between 1850 and 1900) due to human activities, primarily fossil fuel burning, which increases heat-trapping greenhouse gas levels in earth's atmosphere. Since the pre-industrial period, human activities are estimated to have increased earth's global average temperature by about 1 degree Celsius (1.8 degrees Fahrenheit), a number that is currently increasing by more than 0.2 degrees Celsius (0.36 degrees Fahrenheit) per decade. The current warming trend is unequivocally the result of human activity since the 1950s and is proceeding at an unprecedented rate over millennia."[14]

The term global warming is not interchangeable with climate change. Climate change is a separate phenomenon best described as a long-term change in the average weather patterns that define the earth's local, regional, and global climates. These changes have a broad range of observed effects synonymous with the term. It may be of some value to look back at the not too distant past if the collective we on earth want to make peace with our planet today.

SEVENTY YEARS OF REGRESSION

It may be valid to assert that some aspects of society were "greener" in the middle of the last century than is the case today. Although industry was not as environmentally aware then, accidents do still happen today despite much more stringent regulations governing industrial waste. Many developing nations have weak environmental regulations, which are often well behind developed countries, primarily due to the prohibitive cost of implementation and the impact on the bottom line. Aside from the obvious that there were fewer people, homes, cars, and electronic devices back then, what I am looking at is the average person and typical home. I haven't done any formal research and am relying solely on my recollections as a young person in the 1960s.

Children were rarely driven to school; it just wasn't a thing. We all walked to elementary school no matter the weather, and in later years when it was too far to walk to the more distant secondary school, we rode our bikes. It was never considered a hardship; it was not that far, maybe a ten-minute walk for a six-year-old and later as a teen, also about ten minutes if you had a decent bike. Compared to today, and although I am sure not all parents drive their children, many parents do, which results in an extra twenty car trips per week for ten months of the year. For some time, traffic congestion has been a significant issue around schools at nine and three. That's a lot of extra fuel consumption and carbon emissions in comparison.

Then again, two-car households were not the norm, plus the sole vehicle wasn't available on weekdays as the wage earner took it to work. It is much different today, where it is not unusual to see four cars parked outside a house if four adults live there. Vehicles on the road back then were not equipped with pollution reduction devices, as is the case today, and it was also the golden age of the "muscle" car powered by large eight cylinder engines. Gas was inexpensive then, at least until the formation of OPEC in the 1970s. What soon followed was the first post-war shock to the global economic system and western lifestyles, indicating that going forward, it would not be business as usual which remains the case today. The energy crisis led to a significant uptake in fuel efficient vehicles most of which were manufactured in Asia. Half a century later, Honda, Toyota, and others continue to be highly regarded for their quality and fuel efficiency. Still, it's interesting to reflect on that period when one might have thought the spike in fuel costs would have kickstarted alternatives such as electric vehicles yet it was four decades later before the EV push began.

Home delivery of most anything is now big business, undoubtedly boosted by everyone staying home during the pandemic. Companies advertise on TV and often feature famous entertainers in humorous bits to pitch why their delivery service is the best. The delivery craze has grown due to a segment of the population that believes they are too busy to figure out how to plan shopping trips or meals more than twenty minutes in advance.

This delivery obsession we are currently immersed in is little more than a glossy solution seeking to solve a problem that doesn't exist. Ordering whatever you want, whenever you want, has become normalized, but beyond cold food and the additional expense it is not exactly environmentally friendly. Ordering delivery numerous times per week stands in stark contrast to making one shopping trip to buy food for the week and making dinner at home. Picking up take out while on your way home is enroute and there are other exceptions, such as densely populated downtown areas where pedal power is often used to deliver orders. At least one pizza company is now advertising the fact they are trying to be "green" by adding electric vehicles to their delivery fleet. It is possible that the high inflation environment of 2023 will see a reduction in this activity as household budgets are challenged.

There are some job titles that no longer exist, such as milkman. I can recall bottles of milk being dropped at the front door a few times a week, and bread as well. It was a thing, basic food necessities, right to your front door, very practical in hindsight. The empty glass bottles were left outside the door for pick up when the next delivery was made. The collected milk bottles were returned to the plant, sanitized, and used repeatedly. Seems pretty "green" in retrospect. In my district, a ten-cent deposit on milk containers was just added. Before that, the empty containers went into the recycling bin; now, they get driven to the deposit return depot for a refund. I don't know exactly what happens to these plastic containers next but am certain they are not washed and refilled.

Clotheslines were common and were used even in the winter. Yes, sometimes the jeans froze. There were dryers then, but as is the case now, they tend to use a great deal of electricity when running and are still cited as the home's number one energy hog unless an EV is charging in the garage. There was no organic section in the store, it just wasn't a thing, and likely for good reason, as vegetable gardens, berry bushes, and fruit trees were planted as a matter of course. From the garden to the table in short order. A resurgence of people planting produce gardens is one positive thing that came out of the pandemic, with many continuing this newly established practice.

Back then little was tossed, items were of a higher quality so repairing things was the norm. The stereo system I purchased in 1975 still works perfectly as do the AKG headphones purchased the following year. Back then if shoes wore out they were repaired, if you gained or lost weight the pants were altered, if a button fell off it was sewn back on. If you needed a costume a relative sewed it. These days consumers routinely purchase low-cost garments that look great until they are washed a few times after which they begin breaking down so they get tossed or donated to make space for a new batch of what are now often referred to as disposable clothes.

When examining power consumption, comparing the modern world of today to the middle of the last century is complex. Back in the 1950s, not all homes even had a television. Today, it is not unusual to find that a family of four with two teenagers will have at least two but may have as many as four, all of which could be on at the same time. Four modern flat screens likely consume less energy than one CRT from the 1950s, but it is also likely that the TVs are turned on a lot more since there are unlimited channels 24 hours a day.

Computers are also plugged in mostly around the clock, and electricity supply companies note all the plugged-in electronics draw energy even when not in use. The number of people living in a house often equals the number of computers, although cell phones have put a dent in that equation over the last decade. There are billions of phones on the planet that all need to be charged on a regular basis, not to mention a greater number of other common power-consuming items that exist today, like microwaves, gaming consoles, and so forth. They either didn't exist or were in their infancy back then. On top of all that, there is the relatively recent development of cryptocurrency which is reported to consume an extreme amount of power in the process of mining crypto.

For a more complete perspective, consideration has to be given to the growth in the global population and the upward movement to middle class that has lead to an increase in the usage of modern tech devices. Still, there may be lessons from the past that can be incorporated into today's world. I am noticing more couples making do with one vehicle once the kids have moved out, in addition to reducing their overall footprint by moving from a three thousand square foot house to accommodation less than half the size. I personally know four couples that have either parked or sold one of their vehicles.

ENVIRONMENTAL COSTS OF CONVENIENCE

Some of the world's largest corporations, generally considered to be retailers, are really in the convenience business. Online ordering has been marketed as an ideal solution for busy people lacking the time to attend a retail store in person. Those that fully embrace point-and-click probably purchase more stuff as the physical relationship between parting with the funds and receiving the goods has been severed. Taking out fifty dollar bills to pay for shoes in a retail store registers differently in the brain than pointing and clicking, where payment is automated.

Online shoppers don't always receive the products they thought they were buying; sizing and quality can be problematic not to mention delayed or missing packages. A new shopping trend has emerged whereby online consumers routinely order more clothing items than they intend to keep preferring to try them on at home before deciding on the keepers. This convenience results in extra trips for the delivery company or consumer to return the unwanted garments. This practice has now become so widespread that a service company was launched in Toronto to manage the process of returning all the unwanted items to save consumers the trouble of doing so themselves. It has several small retail locations around the city where consumers can drop off the items to be returned.

Excess packaging is an issue as oversized boxes seem to be the norm. I know; I see them arrive, as others in the house sometimes place online orders. I can't help but notice these are brand new boxes that could be reused instead of filling up the recycling bin. By not retrieving their waste material, these companies are passing on costs to cities and municipalities that must deal with this cardboard and packing material. It's time for these companies to go "green" and take back their perfectly reusable boxes. Having an all electric delivery fleet would also be a step forward but EVs still cost more than gas powered vehicles and range is an ever present issue.

It is not clear why it would be difficult for these companies to ship already boxed items as is. When an additional box is needed, a simple retrieval scheme could save billions of boxes from recycling if those left previously were collected during a subsequent delivery. It might be the case that companies are finally becoming aware as a package delivered last summer had a sticker on it noting the outer packaging was lighter than their smallest box.

Another sticker noted the bag was made from 50 percent recycled material and was recyclable. The main thing that struck me though was the strong, peculiar synthetic chemical smell emanating from the bag. I don't recall ever smelling anything like that from a plastic or reusable bag. In mid August 2023, the biggest delivery retailer announced it intends to reduce the use of cardboard boxes and will test no box delivery while also suggesting their sellers use packaging that is ready to ship as is.

For now though, Decycling these boxes would be as simple as copying the milk delivery example from the 1960s. I'm sure the drivers are robust enough to wrestle an armful of flattened boxes back to their van. If enough people started placing the flattened boxes from previous deliveries at their front door, companies might take the hint. Companies could even facilitate the returning of boxes by offering a nominal credit. Avid online shoppers would appreciate it since they would save time ridding their homes of this extra cardboard. For companies, a cost reduction in the procurement of new boxes should be sufficient motivation to collect the boxes for reuse.

Of all the items delivered to homes in a cardboard box perhaps nothing is more inefficient and wasteful than the pizza box, a cumbersome box used once for 30 minutes. Considering we are now well into the 21st century with an ever pressing need to be greener it's clearly a big miss. One solution is a direct handoff. The insulated warmers the pizzas already arrive in could be passed to the consumer who would then remove the pizza in their kitchen—no box required. For outdoor occasions, the answer is a reusable box.

PizzaRound has developed a plant-based compostable pizza box made from renewable fibres. This is a timely development and hopefully will prove to be a cost-effective alternative; however, the box does end up in the landfill, albeit on the compost side. Some may think it's just pizza; what's the big deal? Truth is that two of the three billion pizzas sold in America each year are transported in these boxes. "That's a staggering number of pizza boxes and even more waste - roughly 674,614,000 lbs. of cardboard boxes. All those pizza boxes are equal to the weight of over 234,000 cars!"[15]

Reusable boxes and no box drop-offs are far better alternatives. Container deposit systems have existed for decades and could easily be applied to reusable pizza containers. Reusable boxes could be washed and sterilized at the store. The first pizza company to introduce environmentally friendly options will gain a first-mover advantage for their brand with competitors likely quick to follow.

RECYCLING CHALLENGES

It isn't reassuring to discover what really happens to a great deal of recycled material. Like most people, I thought that the material I cleaned, sorted, flattened, organized, and placed into the recycling bin was processed into a new product or packaging, however, not all the material collected at curbside gets recycled. Researchers estimate that only nine percent of all plastic waste in the earth's history has ever been recycled.

In 2021 the United States exported 74,000 shipping containers of plastic waste overseas on carbon-emitting ships. For many years China was the primary destination for the world's plastic waste, but this has recently shifted to the Philippines, and some Latin American countries as China finally decided no more. Most of this shipped plastic waste is either put into a landfill or burned, with some dumped into rivers which is far from being environmentally friendly. On top of that, the material is transported by fossil fuel-burning ships.[16]

This grim record is an affront to the majority of consumers that put in the extra effort to clean the plastic materials as directed prior to placement in the recycling bin. Even worse, hot water is required to effectively clean some of the empty plastic containers, although going to these lengths evidently doesn't matter. "Canadians throw away over 3 million tonnes of plastic waste every year. Only 9% is recycled while the rest ends up in our landfills, waste to energy facilities or the environment."[17]

Sometimes recycled material is incinerated, which drastically impacts air quality near the incinerator. A recent and rather startling video on the YouTube platform documents the extensive travels recycled material takes. It dispels some of the publicity around recycling and clearly shows that all is not as we believe.

"Bloomberg placed trackers in three plastic bags and dropped them in Tesco supermarket recycling bins around London, England. Their journey revealed a netherworld of contractors, brokers and exporters that looks less like the circular economy and more like a game of pass-the-buck."[18] The resulting travels were not only costly but even worse had associated carbon emissions each step of the way. A posted comment under the video notes that plastic material is being illegally burned in his neighbourhood every night poisoning the air, land and water.

The recycling industry complex and governments at all levels prefer the general population doesn't find out their efforts are wasted out of concern people will abandon the recycling mindset fostered over the past several decades. A waste services brochure recently arrived in the mail containing an impressive amount of information with instructions on what goes where. It noted that plastics from Vancouver Island are carried by ship to the mainland, where they are recycled into pellets for use in other products.

Recycling programs are a net expense to towns, cities, and municipalities as a large amount of what is collected remains a garbage liability they pay to get rid of. The costs for recycling are higher than most people think, while some likely believe that all recycled items have value. In February of 2023, an article was published about this. "The reality is that expenses associated with recycling costs taxpayers millions every year, last year they spent over $168 million to cover costs that the large corporations who produce this waste should ultimately be paying."[19] The cost cited is just for the province of Ontario. Sixty percent of Canada's population is located outside Ontario, so for all of Canada, that figure is closer to $400 million. "Eight million tonnes of plastic end up in the ocean each year. Eleven percent of all the methane in the atmosphere already comes from trash, and methane is eighty-eight times stronger than CO_2 for global warming. So, clearly, the recycling model has failed and needs a complete revamp."[20]

To make recycling more efficient would entail having everything other than garbage and food waste, placed in one container which would eliminate all the driving required to get specific items to the designated drop off locations. Some European countries already have more streamlined systems in place. The recycling brochure I referenced did confirm the need to make several car trips highlighted in the "Take to Drop Off Depots" section which listed nearly a dozen items that all had a specific place to go. I reside in what is considered to be a relatively eco-friendly region, yet residents are expected to put in a lot of time and expense driving all over the district to take part in saving the environment. Some residents, concerned about all the driving, have written letters to the editor of the local paper noting how ridiculous it is, stating they would not be participating.

Returning containers to obtain the deposits has devolved into an arduous task. The main deposit return depot is a 24 kilometer roundtrip drive, with the whole procedure taking at least an hour. Then there is the cost of fuel to make the trip on top. A small satellite depot is closer, but it's rarely open and only allows two people inside at a time. It takes far too long to accomplish such a simple task. Once you get to the main depot, there are at least a half dozen steps involved prior to obtaining the refund. Between the waits, the sorting, getting soaked, or dodging wasps, it's a miserable experience.

I have spoken with others about this depot and they all say the same thing; it takes too long and is not worth the effort so they avoid it and instead throw the deposit return items into their recycling bins. This "system" is yet another example of bureaucratic nonsense enacted under the environment saving banner that is so hopelessly inefficient in practice that it actually discourages people from doing the right thing. I've been to the main depot twice this year, both times during a weekday and it was closed so I donated the returns but was agitated, all that driving and a big waste of time.

There is a TV ad running that perfectly illustrates the bureaucratic detachment from reality. The advertisement informs the viewer that it is now easier than ever to return containers for the deposit refund. After creating an account, you are able to drop the items off at newly created drop points and go. Then later, your account gets credited. So if it can be done, then why not just place them in the recycling bin at home and save all the driving expense, emissions and time? The most reality-detached aspect of the cartoon style ad was how it showed a smiling person in their convertible complete with exhaust coming out of the tailpipe as they merrily drove across town to the newly created drop point!

The solutions are hardly complex. One option is to remove the deposit fee and collect the items in general recycling. Perhaps a better option is to keep the deposit fee, close the return it centres which would stop all the driving, and instead place the items in a deposit-only bin at home or into general recycling. The collection trucks are already stopping to pick up food waste, recycling, and garbage, so adding one more bin would eliminate a lot of emissions from driving. Credits from the deposits could either go to the city or, with a little effort, attributed to each household. I really doubt too many actually care about forgoing a few dollars a month if the items were placed into general recycling, most would be happy to skip the driving and the ordeal to return these items. It was interesting to see a brand new collection bin outside a private liquor store labelled, donate your empties to charity. No extra trips, each time a consumer shops there, they simply pack along their empties for disposal, and for a worthy cause.

A bureaucracy serves only itself; logic is extinguished. The only thing these faceless entities do consistently is expand. When the last global climate conference was held, several newscasts ran video showing row after row of jets that had arrived for the event. Most, if not all, had transported state officials, dignitaries, and their entourages so they could meet, give speeches and agree on actions to combat global warming. The reporting highlighted the hypocritical nature of all these aircraft flying thousands of miles spewing carbon into the atmosphere in order to come to an agreement on how not to have carbon go into the atmosphere.

A HYPOCRITICALLY DIRE SITUATION

It's damaging to credibility on the fundamental issues surrounding global warming when examples are shown where "certain" individuals emit carbon. People conclude it actually doesn't matter for some or for anyone or may think maybe they know something the rest of the world doesn't. I do wonder at times, given what transpired during the pandemic lockdowns, when many were reluctant to enter a grocery store for fear of becoming deathly ill. At the same time, some political figures were evidently unconcerned and were caught taking trips or partying while everyone else stayed home and followed the health directives. The hypocritical do as I say, not as I do on full display.

Is it plausible the overarching measures that shut the world down for a time had everything to do with supressing emissions and little to do with an easily spread virus and was simply a saleable cover to reduce emissions on a global scale? Videos of smog disappearing in highly populated cities as traffic emissions fell were widely circulated. Other videos showed wildlife and fish returning to places they had not been seen in decades. Those were reassuring images; however, the overall emissions decrease is still being studied but is likely not that significant as people were at home more than would have been the case and so used a greater amount of emission generating energy for lighting and heating.

In late 2022 I watched an interview featuring Al Gore discussing the COP conference that had just ended. He spoke about mitigating global warming and explained that technology and satellites can accurately determine where emissions are coming from. "We have started with the largest 72,000 plus, the largest emissions sites around the world. By next year, we'll have millions. And essentially all of them, we won't have the backyard barbecues. There's a long tale. But we'll have 99 percent of the emissions identified where they're coming from."[21] In the ensuing discussion, he noted they only supply the information like a neighborhood watch program; enforcement would be up to each country, city, and state. He's really just talking about a mechanism to ensure accountability however the way technology is being used is unnerving.

3

A WORLD IN TRANSITION

ELECTRIC AIR TRAVEL

It was recently reported that Air Canada was buying thirty battery-powered electric planes from Heart Aerospace to enable the company to lower emissions and fuel costs. These planes are destined for use on shorter regional routes and can carry a moderate number of passengers. "Sweden based Heart Aerospace electric-hybrid aircraft will have capacity for up to 30 passengers and generate zero emissions when they enter into service, expected in 2028."[22]

In late 2022 Canada's first and only type of certified electric plane was delivered to the Waterloo, Ontario, airport. The aircraft was manufactured in Europe by Pipistrel and is the first certified all-electric plane in the world. Usage guidelines indicate that one hour of charging is required for every hour in the air so this plane can only travel short distances. Prevailing weather conditions and airspeed are factors in the distance it can travel.

All electric air travel viability comes down to weight ratio issues. In the future, as commercial electric aviation comes online, the price for these seats will likely be higher than is the case now since the added weight of the battery systems has to be offset by a lower passenger count. The critical developments for aviation efficiency revolve around much lighter and likely smaller batteries which rest on continued technological developments in battery capabilities. There is some discussion that a new battery design with a million mile life span that is no more costly than those currently available is showing potential. The step forward here has to do with how long the battery will function effectively which looks to be much longer than is currently the case. I am sure some will be excited to be among the first to try all electric air travel but for many, including me, there will be a wait and see period, weather conditions can change unexpectedly.

ELECTRIC VEHICLES

The most sceptical automotive enthusiasts view electric cars as experimental due to their lack of history and the fact that it may not be as simple as replacing Internal Combustion Engine (ICE) powered vehicles and carrying on. Many countries, provinces, and states offer substantial rebates to entice buyers. This is a key factor in uptake as it has now become clear that in jurisdictions where subsidies were removed, EV uptake stagnated or fell. The irony with these rebates is that lower-income taxpayers are in effect subsidizing the more affluent as the primary cohort that can currently afford an EV. As well, road taxes are collected when filling up at gas stations but are not when charging up an EV. Some governments are now looking at this and considering a road usage levy to level the playing field. The idea is to base it on distance, the more you drive, the more you pay which would apply equally to gas and electric powered vehicles. Electric vehicles are still more expensive than ICE vehicles; however, this premium is shrinking as increased competition is now driving prices down. Recently, the leading EV maker cut prices twice in an attempt to retain the number one position in market share as competition from legacy auto companies ramps up. Increased EV uptake and growth has to play out in the real world where scarcity of battery input materials is forecast to be an ongoing constraint.

Cheaper entry-level electric vehicles have less range which may render them unsuitable for extreme winter conditions and in rural areas. These cheaper EV models are suitable for drivers primarily taking shorter trips inside a metropolitan area. Much like gas combustion engine ratings, where the manufacturer states the expected gas mileage based on testing in ideal conditions, ranges for electric vehicles are also similarly overstated and of course as the battery degrades, will slowly decline over time.

First, manufacturers caution against draining the battery entirely and strongly suggest charging when the indicator shows 20 percent battery life to prevent premature battery degradation. Second, are the environmental factors when operating an EV. Driving on a 15-degree Celsius windless cloudy day won't require the use of power draining accessories that reduce overall vehicle range. The effects of extreme temperatures, on either side of the spectrum have now been documented in real-world conditions as a factor that can impact range. EV owners that turn in their electric vehicles and return to gas cite limited range and lack of easy charging when on the road as primary reasons. More chargers are being installed and range is increasing but as of 2023, these continue to be real world factors that are impacting buying decisions.

As electric passenger trucks now come on the market, towing with an EV is being examined. Warnings are emerging that the added weight can produce too much kinetic energy and negatively impact the braking and electrical systems. Another well-documented downside for EVs is the propensity to spontaneously catch fire, even when not being driven; although some say this is overblown by the media as gas vehicles also catch on fire. There are however at least two differences, when an EV catches fire, it can be difficult to fully extinguish, and in a few cases, even after the fire was out, and the car was towed away, it burst into flames a second time. As well, ICE vehicles don't spontaneously combust when turned off while this phenomenon does take place with EV's. One car manufacturer recently issued a warning to owners of their EV offerings not to charge them inside the garage overnight due to the possibility of fire. Battery combustibility has also been in the news in relation to electric bikes. Apartment buildings have burned down after catching fire from the battery attached to the bicycles stored in the underground parking area.

Overall practicality of EV ownership is increasingly in focus as adoption grows. The TV ads tend to be homogenous in showing a single-family home with a garage where their electric vehicle can be easily charged overnight. For EV owners without a garage, home charging may not be as painless especially for owners living in apartments and parking on the street. Those EV owners have to rely on the existence of a robust charging network with chargers that both fit their car and work or run a cable out to it and hope it is there in the morning. Even if a building has an underground parkade with a few charging stations, it could be a logistical nightmare to get access when needed. The building owner or the strata could agree to install chargers at every parking spot, but that would be costly with older buildings requiring an expensive electrical upgrade first. New condominium projects often come with chargers in place.

Commercial tractor-trailers have been in development for a few years and are now starting to hit the market.

Some industry experts believe hydrogen is the best "green" solution for big rigs. Currently, Tesla and legacy companies such as Volvo, have consumer trucks either available or close with a driving range of up to five hundred miles or eight hundred kilometres. Tractor trailers for commercial goods transport face similar issues as all-electric air travel; the battery weight offset displacement reduces the amount of goods that can be carried. Also, highway charging infrastructure needs to be accommodative for long-haul distances.

Despite the challenges, as is always the case with technological innovation, engineers will continue to make advancements and bring increasingly better products to the market. One example of this type of progress is the recent development where the battery has been engineered into the vehicle's structure to reduce overall weight. Essentially, the battery does double duty as the power source and as part of the vehicle shell. Production is anticipated to be simpler and cheaper, with range extension the anticipated payoff as overall vehicle weight is reduced.

As of early 2023, several countries have announced dates after which only electric vehicles would be permitted for sale but cracks in these plans are appearing as Germany is now wanting a loophole for ICE vehicles running on "green" fuel. The biggest issue of all now coming to the forefront has to do with delivering all the electricity that will be required. The existing electrical grids were built over one hundred years ago and will need significant upgrades if this transition is going to unfold seamlessly. In the United States alone, the length of the electrical grid is close to 200,000 kilometres, or just about half the distance to the moon. The necessary grid upgrades will be expensive and will require substantial amounts of mined material.

As advancements continue in the coming years, zero emissions transport of all types will eventually become the norm. Using electric transport exclusively to be "green" requires the source of electricity to also be "green." It is counterintuitive to drive an electric vehicle to reduce emissions if the electricity used for charging it is generated from a high emissions source like coal or wood. Some studies on this issue claim it is still better for the environment to drive an EV, even if it is charged from power derived from coal plants.

Other studies indicate it is better for the environment overall to drive a tiny gas-powered car than a giant all-electric hummer. None of this is simple, but the transition to all "green" energy is a megatrend that will continue, bumps and all. New tech-related "green" developments generally come with a steep price tag, so it is increasingly apparent some effort should be focussed on improving the efficiency of existing systems to reduce pollution and emissions. If costs were not an issue, global warming would have been solved many years ago.

ENERGY STORAGE CHALLENGES AND ADVANCEMENTS

Manufacturing the lithium-ion batteries most commonly used in 2023 is a complex process. Ethically sourcing the various input materials, such as lithium, nickel, and cobalt, remains challenging. Of these, cobalt is particularly problematic as the majority used in global manufacturing for EV's, phones and other battery powered devices comes from the Democratic Republic of Congo or DRC where horrific mining practices continue to be well documented. Redesigning batteries to remove cobalt is one solution that would theoretically stall these practices; however, to date, cobalt has been a key component for range and battery stability. Cobalt also has other important uses such as in military hardware and in radiation therapy. One major cell phone manufacturer is looking to distance itself from the ethical cobalt sourcing challenge by stating it intends to use only recycled cobalt in their products.

The Chinese are now ramping up production and usage of a battery that does not contain nickel or cobalt called a LFP (lithium-iron-phosphate) battery and are using them in vehicles, some of which have a 600 kilometre range. The most popular EV in China sells for less than $5,000 US but the base model only has a top speed of 100 kph and a range of only 120 kilometres. Toyota recently announced intentions to produce vehicles using much improved battery technology and is expecting range will be nearly double that of today's electric vehicles. Advancements in battery technology are coming rapidly. In China, new electric cars scheduled for release by the end of 2023 will be using sodium ion batteries. Indications are these batteries will be cheaper than the traditional lithium ion cell chemistries.

For now though, range will continue to be an issue for potential buyers as a caller to a radio show noted in 2022. He said that overall his EV experience was positive; the 300-mile, 500-kilometre range was good. When asked if all his cars were EV's, he said they kept their gas SUV for longer road trips to avoid any chance of being stranded. I have liked the idea of having an EV for some time, regular gas is now two dollars per litre. Two years ago, while perusing a very sleek looking brand new EV the salesperson severely dampened my interest stating, "you don't want one of those glorified golf carts, the range is so low it is basically useless around here." He then redirected me to hybrid vehicles and began his pitch for those stating this was the way to go.

Some analysts and observers have put forth the argument that electric vehicles are not nearly as "green" as advertised, given that it can take upwards of 500,000 pounds of mined material to produce just one battery. Battery manufacturing is a comprehensive process but can be simplified into; mining material -> processing material -> manufacturing batteries. Since carbon emissions are produced at each stage, the "green" aspect of the EV driving experience encompasses a transfer of emissions from the driver to the locations around the world where the ore was mined, the battery was manufactured, and the car was assembled. Although these vehicles don't emit pollutants during usage, their production results in significant carbon emissions and ecological devastation, with developing nations often bearing the brunt of this pollution.

"This means that accessing about 90,000 pounds of ore requires digging and moving between 200,000 and over 1,500,000 pounds of earth—a rough average of more than 500,000 pounds per battery. The precise number will vary for different battery chemistry formulations, and because different regions have widely variable ore grades. It bears noting that this total material footprint does not include the large quantities of materials and chemicals used to process and refine all the various ores."[23]

At first glance, the claim that a single 1,000-pound EV battery requires 500,000 pounds of raw material does seem too high. According to AFP Fact Check's headline on this 500,000-pound assertion, it is false; however, in the body of the article the review of the methodology used indicated it was not that clear cut. Headlines can be misleading, whether by accident or design. In the discussion section Marko Paakkinen, the research team leader at VTT Technical Research Centre of Finland, judged the figures "roughly correct" but said some appeared to have been rounded up and are "gross generalizations." The only specifics cited though had to do with the amount of copper, equivalent to about ninety kilograms, which seemed too high.[24]

The number of pounds of raw ore is difficult to determine as there are several types of batteries and mining, varying ore grades, and some material could be a by-product of other minerals being mined. What needs to be added to any calculations, though, are the emissions from the thousands of exploration companies drilling properties all over the world in search of economic battery metal deposits. For every thousand exploration companies actively exploring, only a few will locate a deposit of sufficient size and high enough grades to warrant mine construction.

All things being equal, the 500,000-pound estimate is likely in the correct magnitude, so by inference, transitioning the millions of vehicles currently operating in Canada would require trillions of pounds of material for processing. It is probable that more material will be needed in the next several decades for Canada to transition to all-electric vehicles than has ever been mined on the earth. Is it impossible? Not necessarily, but it will require a massive amount of capital, take decades plus there will be carbon emissions every step of the way. Cost estimates are continually evolving with recent forecasts suggesting investments totalling a half trillion dollars globally to meet anticipated demand by 2030. On the plus side, western governments have been announcing large investments into mining and battery manufacturing indicating some understanding of the challenges ahead.

In Canada, a 10-year-old entry level EV can typically be purchased for less than $15,000; however, after a decade, the battery health statistics show the range can be down to just 60 kilometres or 36 miles. By the time the expense of a new battery is added in, going with a brand new entry level EV with better technology is a more pragmatic choice. Although new EV prices are trending down, as of early 2023, electric vehicles still sell at a premium even with financial incentives. When EV prices are equal to or less than gas powered vehicles, range is comparable, and chargers are everywhere, nothing will stand in the way of extremely robust EV uptake. However, that appears to be a few years away.

EV battery recycling is ramping up, with several companies focussed on the viability of capturing these metals for reuse but this is not yet a widespread practice, cost efficiency in metal recovery are still poor. Some have noted that instead of disposing the batteries, repurposing them could be a viable solution since some storage capacity remains. They may work well enough to store energy generated from home-based wind or solar capture systems.

Finding EV battery metals in countries with strong environmental standards happens but the path from discovery to mining can take a decade or more likely two. Searching for and defining economic deposits tends to be a protracted, difficult exercise that continuously requires substantial sums of capital, an ever-present challenge for companies in the exploration field. As well, scarcity with increasing demand is inflationary with China also in the mix on this point. Some analysts believe that when a specific battery metal dramatically rises in price, they flood the market to drive prices back down. Doing so dampens mining activity in the west which stops or delays new supplies of the metal. As of July 2023 I remain involved with two companies exploring for battery metals and so have a close up view of the challenges and can state there is nothing easy about finding sizeable economic metal deposits. I will add that the geologists, technologists and company principals are among the most resourceful and determined individuals I have ever met.

HYDROGEN POWER

Hydrogen-powered transport has potential and is currently in use but is not widespread. A seminar I attended in 2020 covered many interesting points about hydrogen adoption. One of the key issues discussed had to do with the need for refuelling infrastructure, and even in a "green" environmentally concerned future focussed city like Vancouver, only one hydrogen refuelling station existed at that time. The primary takeaway was that hydrogen is well-suited for big rigs and heavier trucks. Batteries are heavy; operating without them increases efficiency, and hydrogen power appears to be a logical fit.

However, hydrogen powered transport remains on the outer edge of adoption. One reason may be the public perception that as a highly flammable gas, it is more dangerous than gas powered or electric vehicles. Additionally, there is an argument that hydrogen is not exclusively "green," as it is primarily produced using natural gas – another example of transporting carbon emissions' source, in this case from "fuel use" to "fuel manufacturing."

In 2022, it was reported that Shell was shutting down their UK hydrogen filling stations citing they were not cost-effective. "H2 filling stations — which each cost at least two million dollars to build according to the Hydrogen Fuel Cell Partnership — are undoubtedly operated at a loss in the UK due to the lack of hydrogen-powered vehicles on the country's roads."[25]

Development options remain open. By 2040 Hyundai envisions hydrogen energy being used in other areas beyond passenger vehicles. The company stated it is more focussed on the commercial sector due to the larger and heavier vehicles currently in use as they emit considerably larger quantities of CO_2. The company has reported some success in the mass production of hydrogen-powered trucks and buses, noting the positive feedback from forty-seven large trucks tested in Switzerland. There are also 193 hydrogen buses currently operating in South Korea, with other countries also expressing interest in hydrogen-powered commercial vehicles.

On the innovation front, VW recently filed a fuel cell patent for a hydrogen-powered vehicle that achieves a range of up to 2,000 kilometres using a system where hydrogen is converted to energy that is used to power an electric motor. This option is expected to be cheaper as it does not require the use of precious metals during production.

SOLAR POWER

Solar power was first used in space in the 1950s. Efficiency has continuously improved since then while costs have continued to decline, especially over the last decade. While solar can be effective in specific usage situations in favourable locations, not all geographical locations will work well year round. Daylight hours typically change with the seasons and wet, cloudy climates have fewer sunny days. Solar installations in these locations produce less energy than those in sunnier regions closer to the equator. Perhaps it is just a matter of time until panel technology advances such that they will work well on dull days. Solar panels are not completely environmentally friendly either as they are manufactured in factories plus end-of-life disposal remains problematic. Although panel life span has improved and is now close to 20 years or more, efficiency drops over time, and eventually they need to be disposed of and replaced.

Emerging start-up companies are tackling the disposal challenge by developing more advanced recycling processes to economically recover the most valuable materials in the spent panels; however, it is not yet that far along. The fact sheet published by the Canadian Renewable Energy Association (CanREA) in April of 2021 states, "The disposal of solar panels is at a low volume today, but this will ramp up in the future, providing greater feed stocks for recycling processes....To put solar waste into perspective, the United Nations Environment Programme estimates that fifty million metric tons of electronic waste (such as computers, televisions and cell phones) are produced annually around the globe. Of this, solar PV represents only 0.5%, or about 250,000 tons annually, according to a 2016 IEA report."[26]

As mentioned, geographic location is critical, with some solar installations that were initially thought to make sense now being scrapped. "The southern Alberta city of Medicine Hat pulled the plug on a $13-million concentrated solar power facility after operating it for about five years....the project's small and unreliable contribution to the community's power needs didn't justify the cost of maintaining its rows of mirrors and pipes through snowy winters and dusty summer days."[27]

It comes down to the return on investment for governments, businesses and individuals, although perhaps less so for those motivated by an overriding desire to be "green," even if the rate of return on solar is, at best, break-even. Several years ago, I experienced corporate decision making firsthand while working with an off-grid solutions company entering the field of solar installations. I was contracted to help with market analysis for their expansion into wind and solar but it quickly became clear the proprietor underestimated the technical expertise required to set up large installations.

A local auto dealership expressed interest in purchasing a solar array as their glass encased showroom used a large amount of electricity in both winter and summer. The owner correctly noted their location was ideal, the building had a flat expansive roof completely open to full sun exposure year-round. Measurements were taken to determine the number of panels it could accommodate, after which power generation potential was estimated.

The dealership owner understood the trade off, he would be forward paying for electricity. The array came in at well over $200,000 not including replacement panels or maintenance. At that time, panels were rated at 15 -20 years and with break-even calculated to be 15 years which factored in higher future energy costs, the project was a non-starter. It was too much money upfront plus a considerable amount of effort for a small unguaranteed potential gain.

A decade later, though, the equation may have tilted in favour of proceeding as carbon credits are now a potential added incentive plus newer panels have a longer life expectancy. Credits could be created based on the amount of power generated that would displace power from the grid. Although the critics of carbon credits and carbon pricing correctly note that the added tax is inflationary, this may be an example where the carbon credits created could have tipped the numbers in favour of proceeding.

The future looks panelled. Eventually, whatever can be, will be covered in high-efficiency solar panels as the world transforms to low and near-zero emissions energy creation step by step. Panel life spans have continued to improve, with some manufacturers now claiming up to 25 years depending on maintenance. Although much of the material that goes into a panel can be recovered, this is an area lagging behind.

"By 2050, the International Renewable Energy Agency projects that up to seventy-eight million metric tons of solar panels will have reached the end of their life, and that the world will be generating about six million metric tons of new solar e-waste annually....There's a cost problem in the US because it's way more expensive to recycle than it is to dump. It costs less than \$1 to dump a panel in a landfill and between \$12-25 to recycle it, but only \$3 is paid out for recovered aluminum, copper, and glass."[28]

TURNING THE TURBINES

Wind and solar power generation share one commonality: both rely on prevailing weather conditions. Some regions of the planet are consistently windy, so are well suited for large turbine installations. A few years ago I authored an in-depth article on a diamond mine located in the far North of Canada. Due to the remote location, there was zero infrastructure, so the only option was to use diesel generators to power operations. To reduce expenses and pollution, "green" alternatives were considered which led to the installation of massive wind turbines at a cost of thirty-one million dollars.

Once in operation, diesel fuel consumption dropped by roughly ten percent, plus the projected life span of the turbines coincided with the forecast life of the mine, so it made sense in that respect. The five million dollars of diesel fuel saved annually meant the cost of the installation would be recovered in 6 years. One thing to note is the favourable location. "South Baffin's near-persistent wind seems to make it an ideal place to take on wind turbine power-generation. The mine's four turbines typically turn to winds averaging about twenty-five kilometres per hour, Bertoli said, and have been powered by speeds as high as fifty-four kilometres an hour."[29]

During the winter of 2013, operational issues were encountered as the turbines froze requiring the installation of heaters to fix the issue. After the heaters were in place, no other significant problems were encountered and in the final analysis the project was considered a tremendous success. This project serves as a solid example of remote wind generation more than paying for itself while reducing fuel consumption and carbon emissions.

Closer to home on a gulf island not far from Vancouver, a homeowner decided to attempt off-grid living even though the grid was available. I heard about this when working with the off-grid equipment seller I noted earlier. The investment was significant at over $50,000, which was the price for one fifty-foot-tall turbine. Islands tend to have winds present much of the time, so to all involved, it seemed like an option that would work well. The decision to go with wind and not solar was due to the densely treed nature of the surrounding land; the forest surrounding the dwelling blocked out the sun even when it was at it's highest point above the horizon.

Even on the windiest days, the turbine proved to be ineffective as those sun blocking trees also screened the wind to a far greater degree than anticipated. The exact location, as with solar, is critical if wind power is to meet expectations but turbine installations do come with other downsides.

According to an article on Forbes.com, "Wind turbines have also emerged as one of the greatest human threats to many species of large, threatened and high-conservation value birds, after habitat loss from agriculture…. The expansion of wind turbines could result in the extinction of the golden eagle in the western United States, where its population is at an unsustainably low level."[30]

As is the case with solar panels, end-of-life protocols remain problematic with recycling options continuing to be under development. Commercial turbines are gigantic pieces of infrastructure, with recent examples reaching three hundred feet in height and weighing as much as eight tons. According to an article written by T. Leonard on wind-watch.org, "the blades [are] built from a 'composite' of fibreglass and resin that can withstand hurricane-force winds [and be] light enough to turn, [so they] cannot easily be crushed, let alone recycled. Scientists are looking for ways to separate the resins from the fibres or grind chunks of blade into small pellets that can be used in other products, but they're struggling to find any process that works on a large scale. Amid the zeal to recycle, there's no small irony in the fact the main sources of renewable energy cannot themselves be renewed when they reach the end of their life. By 2050, it's predicted that the world will need to dispose of two million tons of wind turbine blade waste every year. In the UK, the volume already exceeds 100,000 tons per year. Currently, the world's decommissioned blades are mostly buried in landfill sites, where they will take centuries to degrade."[31]

Technological advancements will continue to address existing shortcomings in all aspects of "green" power generation. A recent press release on testing new turbines is another example of this progress. "Siemens Gamesa announced Monday that its breakthrough development in offshore wind turbine technology, the 14-222 DD offshore prototype has set a world record for the most energy generated over 24 hours. One of the keys to the 14-222 DD offshore prototype's success are its 'revolutionary' blades cast from a single, gigantic piece of recyclable resin."[32] There was no mention of the tower being eco-friendly, though. Nothing in the "green" energy complex is perfect yet.

ENERGY FROM BURNING WOOD

Burning wood to create energy is not considered to be carbon neutral by all scientists and experts. For wood to qualify for the carbon neutral label, a seedling must be planted to replace each tree cut down and burned to create energy. The reasoning presented in the argument for neutrality is that over time, the seedling will theoretically "re-absorb" the carbon released when the mature tree is incinerated. A hidden very destructive downside also exists with this carbon neutral label as it has created an incentive to remove forests for energy which is still happening.

Intuitively, the idea of cutting down and burning mature trees to save the planet seems off base. Basic logic demonstrates the folly of this idea. If it takes one hundred years to grow a good-sized tree, it was during that century of growth the tree was actively removing carbon dioxide. For a 100-year-old tree standing today, carbon removal would have begun around 1920. The disregarded fact is that all the carbon the tree removed gets released back into the atmosphere in a matter of hours today, which is in direct opposition of the stated reduction goal. If the carbon-neutral argument were wholly coherent the planet could be saved if all the trees were cut down and burned for energy as long as seedlings were planted for each one removed. Right.

As an interesting comparative, if every square mile of barren land on earth could miraculously be covered in solar panels tomorrow, the amount of electricity generated would be magnitudes beyond sufficient for the earth's needs. "Deforestation from human activities has resulted in a total of about 180 billion metric tons of carbon released to the atmosphere since 1750 (the approximate start of the industrial era). Plants and soils have reabsorbed about eighty billion tons since that time, with a net historical loss of about one hundred million tons of carbon to the atmosphere through deforestation."[33]

As noted, the case for the carbon-neutral label hinges on planting replacements but a recent study concluded that half the trees planted don't survive, according to tree survival data from 176 sites in Asia. On average, about 18 percent of newly planted trees die in year one, with nearly half perishing by year five. Variations in survival rates depended on the sites and types of trees with newly planted trees in completely deforested areas having the worst rates of survival, which is unsurprising.[34]

Along with inefficiency, wood burning is exceptionally unpleasant, my neighbourhood has homes that still use wood stoves for heating and I have been left choking for air when outside in the cooler months. Local governments continue to debate banning them due to the massive degradation in air quality when in use. I don't think people need convincing. It is June 2023 as I edit this section. Right now, newscasts are showing smoke from wildfires covering large areas of Canada and the northern United States which is causing disruptions on a number of fronts including air quality health warnings.

GASOLINE AND NATURAL GAS POWER

Common types of gases used worldwide include diesel, gasoline, and natural gas. Natural and refined gas emit just over 200 grams of CO_2 per kilowatt hour of energy, with gasoline, oil, and diesel higher at roughly 265. Coal is much worse at 338, then worst of all is wood at 368; however, that is non-sustainable wood but, in reality, is still wood.[35]

Alternatively, Natural Gas has many applications, including heating, cooking, creating electricity, and as a transportation fuel. Natural gas is the best or least damaging of the undesirables, so until the earth is running on 100 percent renewable energy, moving away from the worst emitters like coal and wood toward cleaner options, is worth doing and will help. Although natural gas is not near zero, emissions are much lower than is the case with wood or coal. The faster the world moves towards viable green solutions, the better; however, all transitions take time; this one is no different, and the need to be rational about the transformation is critical.

I suspect one of the reasons some still push back against "green" energy is the at-all-costs mentality aligned with "green" mandates that often come right from the top. It is important for everyone to engage in solutions regardless of whatever rationalizations may be conjured up that serve to convince the individual that those efforts are too insignificant and won't matter. A sober second look is needed for those that want to wash their hands of it by suggesting the science is flawed or global warming is a hoax. The primary points to consider are that we don't know how it will play out but do know that any and all pollution is terrible; the less we pollute the water, land, and air, the better for every living thing on earth. This rudimentary truth may have been lost in all the posturing and shouting. The future is unknowable, the warming may not be as dire as some are predicting but it's clear that the path forward is to reduce pollution of all types. The old saying comes to mind, prepare for the worst but hope for the best. Speaking of the worst, it is now August 2023 as I wrap up final edits and related news just keeps on coming with new fires being reported daily. From local fires in BC that have displaced residents in several towns to the highly destructive and deadly fires on Maui. At the same time, the first hurricane to hit California in 84 years has resulted in widespread flooding.

HYDRO ELECTRIC POWER

Nearly 90 percent of the energy consumed in my home province of BC is from hydroelectricity produced by turbines embedded in massive dams. Surprisingly, solar and wind combined comprise less than three percent of the energy used in the province. By inference, electric vehicles operating here are considered to be very "green." Hydropower has been labelled as a renewable energy source with benefits including easy integration with other forms of electrical power, the ability to meet peak demand periods and is a lower cost energy source over the long run.

Dam construction is the most contentious aspect of this type of power generation. The drawbacks include prohibitive upfront costs to build, reservoir availability and lasting environmental impacts from flooding the area behind the dam. Downstream issues include silt build-up and bank erosion with a corresponding negative impact on fish and wildlife. When the area behind a new dam is flooded it becomes a lake. The first time I ventured into one of those lakes, not knowing it was a flooded area behind a dam, I freaked out momentarily as my foot got caught under a tree root, it felt like something was grabbing me from below the surface. What appeared to be a pristine mountain lake was in fact a tree stump graveyard with the stumps near the shoreline sitting just a few feet below the surface of the water.

The most significant potential negative with any massive dam structure is failure. Dams built out of concrete generally don't fail as they are engineered from the design stage to withstand earthquakes, but when a failure occurs, the consequences are often devastating. Globally, dam collapses have resulted in more immediate casualties than from any other type of power generation. Some observers are now sounding the alarm and insisting that hydro electric dam projects are not as "green" as previously thought.

As evidence, the critics note that the entire dam-building process produces massive carbon emissions that have been conveniently overlooked. The amount of concrete used for the biggest dams is staggering. The site C dam, currently under construction in BC, has already used over 1.7 million cubic metres of concrete just for the buttresses designed to ensure structural integrity in the event of a major earthquake. Estimates of the amount of carbon dioxide associated with one cubic metre of concrete vary but can be as high as 40 kilograms per cubic metre although some researchers suggest it is even higher.

TIDAL POWER – CONSISTENT NOT CONSTANT

Harnessing power from the tide is a more recent power generation advancement. The earth's largest tidal power station is situated off the west coast of South Korea and consists of ten water turbines that produce enough power for a city of a half million people. The amount of power generated is roughly equivalent to 850,000 barrels of oil which can be stated as over 300,000 tons of CO_2 not entering the atmosphere. Energy is generated when water flows in and out of the floodgate and waterwheel located between the ocean and lake. Studies done on this installation after it was operating showed that the water quality in the lake was not negatively impacted but had improved due to the increase in circulation resulting from the project.

On the plus side, energy output is more predictable than wind or solar as the tidal movements are known in advance which facilitates the design of more efficient systems. As tidal movements are a consequence of gravitational forces exerted by the moon and sun, this type of power creation might be the most space age of all since power is being harnessed from gravitational forces from the solar system. Other positive aspects of tidal power include a decrease in coastline flooding and length of service, tidal projects tend to last longer than other renewable options.

Upfront costs are considerable including comprehensive environmental studies that must be conducted before moving forward. Ensuring there will be zero environmental impact on marine life and ecosystems is a critical first step. After installation, depending on the location, the equipment can be difficult to repair or even access for routine maintenance. Energy storage capacity also remains challenging as the tides are predictable but not constant. These projects are not simple.

A tidal project in Nova Scotia was just abandoned after years of work and an investment of tens of millions of government dollars. Like many of the "green" energy options already covered, tidal power will become more efficient as designs and engineering continue to advance, however it is likely to remain one of the more challenging "green" energy generators.

RIVER POWER

There are smaller power generation projects that use water flow to create energy now operating with others in the development stage. The Canadian Hydrokinetic Turbine Test Centre is running a development project near Pinawa Manitoba. The project involves placing a turbine into the bed of a flowing river where the current is strong enough to turn the turbine as water flows through. With a strong enough current, the project will be able to generate a moderate amount of energy with little environmental disruption. The expected end users of the power generated will be isolated

First Nations communities that currently create power via generators running on gas and diesel.

Run-of-the-river projects are similar to other small scale hydro generation projects already in operation. "Large-scale hydro plants can cause significant damage to local ecosystems, whereas run-of-the-river projects utilize existing waterways and gravity."[36] Currently, at least two projects of this type are operating on the west coast of British Columbia. One is located near Sechelt, creating 33 MW of hydroelectric power as a partnership between Sechelt Nation and BluEarth Renewables.

The other project, located near Squamish, creates 15 MW and produces power for the local First Nations. They are, in turn, selling that power to BC Hydro, the primary power supplier to the Province of BC. Electrical output is created by diverting downstream water flow into pipes with turbines connected to a generator. After the water passes through the system, it's returned to the stream from which it was diverted. Project cost payback is estimated at 40 years, after which the Squamish First Nations will hold 100% ownership.

NUCLEAR POWER

I can still vividly recall the exact moment in 2011 when I heard the news about the most recent major nuclear accident. The accident I am referring to is the Fukushima reactor in Japan, caused by powerful forces outside human control, earthquakes, and tsunamis. There have been other well-documented nuclear incidents, and it is the lingering disturbing images that continue to serve as the main driver of negative sentiment around embracing nuclear as a viable clean energy solution today.

The tragedy in Japan was not the direct result of the 9.1 magnitude earthquake; rather, the plant melted down due to the tsunami that resulted. Nuclear plants are designed to withstand powerful quakes, as was the case at Fukushima, however, due to the ocean-side location, the resulting tsunami caused the worst nuclear disaster since Chernobyl. The facility withstood the force of the quake but the ensuing tsunami was twice the size the planning design accounted for. The plants' seawater pumps that function to keep the reactor's temperatures down, failed when water rushed on shore. This resulted in the reactors leaking radioactive material into the Pacific Ocean. It's now over a decade later and work continues; it is expected to take up to 40 years to finish decommissioning the plant. Even so, as of 2023, Japan has returned its clean energy needs focus to nuclear despite this relatively recent tragic event.

In exceptional circumstances, there can be long-lasting negative consequences to both the environment and humans when nuclear accidents occur, however, nuclear energy ranks at the top of cleaner energy options due to the absence of emissions during its production. As is most always the case which should be obvious by now, nothing has a zero carbon footprint, and nuclear power generation is no exception. The carbon footprint comes from the materials used in plant construction and mining. Compared to the total amount of clean power produced, the footprint is considered small.

"Unlike fossil fuel-fired power plants, nuclear reactors do not produce air pollution or carbon dioxide while operating. However, the processes for mining and refining uranium ore and making reactor fuel all require substantial amounts of energy. Nuclear power plants also have substantial amounts of metal and concrete, which require substantial amounts of energy to manufacture. If fossil fuels are used for mining and refining uranium ore, or if fossil fuels are used when constructing the nuclear power plant, then the emissions from burning those fuels could be associated with the electricity that nuclear power plants generate."[37]

A former colleague now residing in Germany explained how committed the country was to implementing "green" solutions. He said there had even been discussions about micro nuclear installations designed for home use to meet the energy needs of that home. I had not heard about anything like this so was surprised such a solution was being considered or even existed.

"Scientists are working on micro-nuclear reactors that are so small they can fit on the back of a truck or a standard 40-foot shipping container, explains Christina Nunez writing for the Argonne National Laboratory. Deliberately small, generating up to 20MW, they could provide zero-carbon power in remote settings or supplement electrical power grid recovery. Another idea would be to locate them on remote highways for re-charging long distance electric trucks."[38]

The ever-increasing urgency for low and zero-carbon energy sources has led to a re-examination of innovative and safer nuclear options in many countries. The war in eastern Europe created a dire situation by disrupting gas and oil supplies from Russia. Other than Ukraine, perhaps no other country was impacted as much as Germany since they were counting on Russian exports to meet their energy needs while continuing their transition to "green" energy.

The positive side of the nuclear debate centres on the fact that nuclear can make a significant long term impact in solving the ongoing need for clean energy. The first nuclear reactors were commissioned back in the 1950s, so it is not a recent innovation or new source of clean energy. Today, roughly ten percent of the world's electricity comes from nuclear power generation. Hydro power is first globally but as of 2023 nuclear is the planet's second-largest low-carbon power source at 28% with 50 countries creating energy from 440 nuclear reactors currently in operation. Another 55 are under construction which will add significant capacity when brought online.

"The uses of nuclear technology extend well beyond the provision of low-carbon energy. It helps control the spread of disease, assists doctors in their diagnosis and treatment of patients, and powers our most ambitious missions to explore space. These varied uses position nuclear technologies at the heart of the world's efforts to achieve sustainable development."[39]

The United States leads all countries in nuclear energy production, followed by China, France, Russia, South Korea, and Canada. Improved performance has been achieved from existing decades-old reactors. Sweden recently announced it was pivoting to increase nuclear power generation, likely spurred on in part by the recent instability in the region however, the report noted that discussions on nuclear had already been taking place. "Sweden now joins other countries in Europe that are turning to nuclear power in response to record high energy prices and fears over the security of key infrastructure. More than eight out of ten, or 84%, want to continue to use nuclear power or, if needed, build more reactors…"[40]

The input mineral uranium is abundant and naturally occurring. It behaves the same in a reactor as it does below the earth's crust; as it decays, it produces heat. Canada is home to one of the largest uranium mines on the planet, which also has high concentrations of uranium as compared to other uranium mines. After it's mined, the uranium is refined and enriched prior to use in the reactors. The resulting product is about the size of a sugar cube and is called a nuclear fuel pellet which contains as much potential energy as one tonne of coal.

In November 2022, the Canadian government seemingly gave a green light to nuclear power generation projects when it introduced a 30% investment tax credit for clean tech that included SMRs' small modular reactors. The nuclear industry believes the Canadian government has sent a strong message it considers nuclear a clean energy and a contributor to a "green" future the same way it views other low-carbon power generation technologies such as solar and wind.

What must be acknowledged is the fact that Canada only contributes a tiny percentage of global emissions today. This means that even if Canada got to zero in a few years, overall global emissions would remain largely unchanged in the face of much more populated countries that dominate the emissions landscape.

4
DECYCLE

ORIGINATION

The first time I Decycled was back in December 2002. I had purchased a battery-powered driveable toy car for my two-year-old son but after trying every conceivable way to get it into my vehicle I paused for a moment to consider other options. It then occurred to me it might fit if I removed it from the oversized box and sure enough, in it went. As well, the massive box would have been a pain to get rid of later at home so gladly left it with the store clerk.

After that, I continued to Decycle excess packaging whenever practical, then in 2004 a store cashier asked me what I was doing. I said that I was Decycling, the term just popped out. In retrospect, as far as I know, that was the origination of the term, at least within the context of meaning the act of discarding recyclable or excess packaging at point of purchase. I just kept on using the terms Decycle or Decycling whenever someone asked me why I was removing excess packaging from my purchased goods and leaving it behind at the store.

When I Decycle at the checkout, I explain that I only want the item I am buying and have no use for all the packaging so if they could put it in their recycling that would be great. If the clerk seems at all puzzled I elaborate and say I came in for the shoes but have no use for the box so I am leaving it here, I don't want to drive it home then drag it out to the curb later. After the brief explanation, they usually say right, that actually makes sense. One time I was cautioned about keeping the box; the clerk informed me that the item could not be returned without it. So far, when Decycling, that has only happened once. At grocery stores, the excess packaging is usually left in a cart outside the store.

The practicality of Decycling became increasingly apparent as the years went by. Having less cardboard to deal with at home saved time and it was much easier to fit everything into the small yellow recycling bag. Until recently, collections allowed one small blue bin for tins and plastics and one plastic sleeve for the paper and cardboard. Even after I was Decycling on a regular basis, the excess cardboard from pizza boxes and online shopping made it nearly impossible to fit it all in the designated bag. Home delivery waste material left behind is another area where the companies involved can reduce the amount of cardboard they leave the consumer to deal with. As noted earlier, at least one company is now looking into the feasibility of getting rid of the extra box.

Now, with the larger recycling bins rated at 200 pounds, when I Decycle the cardboard and leave it with the retailer, the bin takes four weeks to fill even with a few delivery boxes in the mix. The recycling bin has been going out every four weeks instead of every two for some time now. Decycling just the cardboard has clearly demonstrated that for a house with four or five residents, the time between pickups can realistically be cut in half and reduced to about once a month from every other week.

A few years ago I realized a notable positive environmental upside was attainable if everyone started Decycling; once collections schedules were reduced, millions of tonnes of emissions would be stopped. A small tweak in refuse management, scaled up, would make a significant impact in many ways. The round-trip excess packaging currently makes is entirely unnecessary. Recyclables currently travel from the store to home, then to the curb, then to heavy trucks for a tour of the neighborhood before being deposited at the transfer depot. By leaving packaging at the point of purchase, the need for transporting the waste material is permanently stopped.

With the recycling bins taking four weeks to fill instead of two, the number of trips emission-spewing trucks make is cut by 25%, and in turn, carbon emissions decrease by the same 25%, in perpetuity. Additional resulting positives include a reduction in truck operation costs for fuel, labour, maintenance, repairs, and replacement vehicles. Secondary spinoffs include less road wear and a reduction in traffic congestion. Once a city calculates the savings realized, council can review options for redirecting those funds. In fact, Decycling could be a full circle solution if the budget savings were set aside and eventually used to purchase electric collection trucks. Having an all electric fleet would close the circle and result in even greater emissions savings for collection activities.

For retailers like grocers, the paper and cardboard Decycled by consumers would be collected in large bins placed either at the front or side of the store. When filled, the bins would be rolled to the back of the store and the cardboard bailed. Most, if not all high-volume retail stores already use these simple machines to compress cardboard. As the bales are produced, they are placed on pallets for pick up by the empty trucks deadheading back to the central warehouse. After unloading pallets of product at the retail store, it is standard practice to fill these now empty trucks with stacks of pallets, bailed cardboard and empty dairy cases before they head back to the warehouse. Moving the discarded cardboard in this way taps into the existing transport infrastructure.

BENEFITS

Decycling began as a personal convenience that resulted in less packaging to deal with at home. Stopping the redundant transport of excess packaging to the home and then to the transfer station is attainable. Noncapital-intensive practical solutions to stop emissions are needed and Decycling is one such solution that is sensible and can be implemented easily and quickly. Overall, solar tends to reduce emissions, as do, hydro, wind, nuclear, and many other "green" initiatives already reviewed; however, all are expensive, not exactly 100 percent "green", and many have end of life disposal challenges. Doing as much as possible to reduce pollution and prevent emissions now should include tweaking existing systems to reduce or stop emissions.

Total emissions stopped becomes highly significant when Decycling implementation scales globally and becomes standardized as an environment-saving practice. The math demonstrating this is covered later in this section but by now, it should be apparent that Decycling is worth doing. It may not be such a foreign concept to retailers either as I am aware of one regional department store chain that encourages their customers to bring in difficult to recycle items.

I have also recently noticed micro recycling drop boxes in a few select retailers but they are only for specific containers from the sponsoring company. Indications are emerging that both retailers and consumer goods suppliers get it and appear to be making an attempt although for some, it may be for appearances sake only. Ads promoting new packaging that reduces waste are now periodically popping up as retailers are increasingly recognizing the need to at least demonstrate they are pro-environment.

With environmental consciousness at an all-time high and continuing to trend in that direction, retailers would be wise to view in-store Decycle points as a positive opportunity. In the early adoption phase, retailers already leaning into some of the initiatives previously noted, will likely be first to implement Decycling and in so doing enhance their brand. As more retailers create Decycle points, the laggards will have no choice but to follow. Cities could mandate it as they are also a beneficiary but this should not be necessary. If it turns out an incentive is needed, it could be structured as tax relief or a one-time grant earmarked for creating a Decycle point in retail stores.

At the same time, it would be an egregious oversight to continue giving big delivery companies a free pass; they need to retrieve their excess cardboard and deal with it internally or stop sending it at all. Doing so would reduce the number of new boxes needed for shipping. I am actually a bit surprised the same drop and run model is still being used.

PRACTICALITY

Not all consumer goods packaging is suitable for Decycling. Although low and near zero waste retailers are now emerging, achieving zero packaging would require a revamp of the current supply and merchandising system. Innovative retail stores are being launched with a near-zero packaging approach aimed at eco-friendly customers willing to fill containers they bring with them. Consumers are already familiar with this concept; most large grocers have a bulk foods section where the labour for packaging shifts to the shopper in return for a lower price.

What these new retailers are trying is commendable and might be a viable shopping alternative for some, but it will be slower. Before heading out to shop, containers must be gathered up and sanitized, then at the store, filling, weighing, and recording the products takes extra time. The added steps and time commitment might be an acceptable trade-off for some but shopping for a large family in these stores would not be as practical due to ever pressing time constraints. While it is true that it does take a small amount of time to Decycle at the retailer, time is saved later as there is less packaging to deal with at home along with fewer recycling cart trips out to the curb and back.

For this new type of retail operation, perhaps the most significant practical challenge involves food integrity from sanitization issues. Having dairy products like sour cream or cottage cheese being doled out of a large wholesale container into one brought from home could lead to food safety questions. A brand new container solves the problem, but using a new container defeats the purpose. In the event a recently purchased food item quickly deteriorated because the container created a science experiment, illness could result if the spoilage went undetected. If premature spoiling did happen, determining which party is responsible is difficult. Longevity guarantees will be practically unavailable on some food items. As it is now, with most major grocery stores, products can usually be returned with little fanfare even without a receipt if they spoil prior to the best before date.

When Decycling, some consumers may want to keep the box to reference the list of ingredients. Knowing the exact number of sugars, carbohydrates, and fats, is critical for some due to health and diet-related reasons. I understand the importance of knowing this information from firsthand experience. At age ten, my youngest son was diagnosed with Type one diabetes, a life threatening autoimmune disease which affects the pancreas. After diagnosis, it was vital to know the exact number of carbohydrates and sugars he was consuming. The easiest way to capture the information was to snap a picture prior to discarding the excess packaging. In time, companies could put this information on the inner packaging as Decycling becomes the norm.

THE EXCESS PACKAGING DECEPTION

At least one major food supplier has already stopped using an outer box on some of their products. The packaging was marketed as new and improved which was true; the package was easier to both open and reseal. The iconic branding was placed on the outside of the new packaging, easily recognizable, no different than before, except the box was gone.

Reducing excess packaging will not be popular with food production companies since oversized boxes effectively hides food inflation. For several years consumers have been aware that both package sizes and the amount of food inside the package has been decreasing. Companies routinely put in less food while keeping prices constant as a way to camouflage rising prices while hoping consumers won't notice. This practice is now so widespread it is being covered by the MSM and given the term shrinkflation.

Not long ago I was more than slightly annoyed when I placed a frozen pizza on top of the box it came in and one-third of what I was expecting to see was missing. The contents of this particular brand used to fully fill the box it came in. Food companies use these tactics hoping to preserve profit margins but shrinking the amount of food inside an opaque box is pure deception as prior to purchase consumers can't see how little is inside. Companies are probably correct to be concerned that customers would pause and reconsider the purchase if they could see how little food they were actually getting if the outer packaging were right sized to the amount of food inside. Companies have little incentive to right size packaging and every reason to see what they can get away with before the complaints become loud enough to force changes.

Consumers are paying the price for the illusion being perpetrated but there are a few exceptions, some companies have decided their brand is strong enough to retain correct-sized packaging. An ice cream company recently ran an advertisement claiming they had not downsized their portions or raised their prices, like their competitors had. Using this claim as a unique selling proposition says a great deal about the current sizing and packaging trends.

Another hidden cost that indirectly impacts the environment negatively is the transport of partially filled packages all over the globe. Assuming a conservative 25 percent over packaging problem, once corrected, would result in transport trucks and ships being able to carry 25 percent more product which would reduce transport costs and emissions accordingly. Overall load weight would still be a restrictive consideration.

Recycling cardboard is one bright spot in the recycling complex. Cardboard recycling saves trees that when left standing continue to remove carbon from the atmosphere. Other positives include water and energy savings, reduced pollution and emissions reduction. Most people that I talk to say they recycle whenever possible including the tiny cardboard toilet paper rolls. If it's cardboard, it goes into recycling, not the trash; it's not hard to do.

A standard bale of cardboard has a dollar value, unlike some other recycled items, where disposal costs are paid by the city to move it along the chain. "If you have a business, like a grocery store, and you use large amounts of cardboard, you could be profiting off your cardboard use."[41] Some businesses like grocers could gain up to $5,000 per year if all their cardboard were compressed and sold. Cardboard can also go through the recycling process numerous times before breaking down, which is another positive for the environment. Typical uses for recycled cardboard other than making more cardboard include paper, paper towels and tissues.

DISRUPTIVE ADJUSTMENT TO AN INEFFICIENT SYSTEM

Huge sums have been spent trying to meet global emissions goals; however, progress has been remarkably slow. "At the end of last year, overall fossil fuels represented 81% of energy consumption, 10 years ago, they were at 82%," says Jeff Currie. "$3.8 trillion of investment in renewables moved fossil fuels from 82% to 81% of the overall energy consumption."[42]

That is a massive expenditure to gain only a one percent reduction. Theoretically, if the figure were 40 trillion then a ten percent reduction would have resulted. Of course the numbers are unknowable but what is known and is a recurring theme is cost. It should be clear by now that more focus needs to be on the implementation of innovative cost-efficient practical solutions like Decycling to permanently stop emissions and reduce pollution. It is not an all encompassing solution; nothing is, however Decycling is a worthwhile initiative that will help and can be implemented now.

For Decycling to become the norm, there must be action. Although local governments will save money it might be overly optimistic to think they will drive it, rather, Decycling will most likely begin as a grassroots consumer-driven effort led by environmentally concerned individuals willing to take the initiative. As more participate and Decycle, retailers will be pushed into taking accommodative steps. I spoke with experts in waste management, and the consensus was that Decycling will become formalized and normalized once consumers exercise their right to leave the excess packaging behind. Cardboard is just the starting point; once established, the focus can shift to plastics. Apparently there are over 20,000 pieces of plastic in the oceans for every person on earth. Chemical recycling is showing promise as a step forward for plastics but is expensive, has high start up costs plus the economics still favour making new plastic materials from fossil fuels.

Many realize more needs to be done. Whenever I've explained Decycling to those under thirty years of age, I tend to get a positive enthusiastic reaction with comments like "cool, that makes sense, I am going to start doing that." The reaction from those middle age and older isn't always the same as the younger cohort, they tend to point out potential challenges and note that retailers may resist. The differing reactions by age makes sense since younger people have the most significant stake in what the planet will look like in the second half of this century. Those over sixty today may have less concern since they won't be around to experience the state of planet earth in the second half of this century although their children probably will be.

Given that the first objective is to reduce overall truck trips by 25%, pickup logistics requires that food waste pickup also be reduced to coincide with the reduction to achieve no pickup in week two. Collections are carried out a bit differently depending on the city or municipality, however, there are no logistical challenges that can't be worked out. There is no downside to reducing food waste pickup frequency, the bin is large enough. The only potential complaint may come from odour in hot weather if anything more than eggshells, coffee grounds and bones are in the food waste bin. In hotter months, the simplest solution is to put food waste in a secure compostable bag and then directly into the freezer. Any smell is stopped once food waste freezes. On the morning of scheduled pick up the bag can be placed in the bin.

SAVINGS AND CREDITS

A carbon credit is a permit that allows the credit owner to emit one tonne of carbon dioxide or the equivalent of other greenhouse gases. Carbon credits are generated when emissions currently being released are stopped and from projects that pull Greenhouse Gases (GHGs) out of the atmosphere. A programme to Decycle meets the definition as it stops emissions that were occurring. Each time a project verifies the reduction, avoidance, or destruction of one metric tonne of GHG, one carbon credit is created. In common American usage the word ton is generally accepted to mean 2,000 pounds. The advantage of using metric tonnes is the consistency across jurisdictions. A metric tonne is 1,000 kilograms or kilos and is used most everywhere except the United States which stands as one of the few countries that has not adopted the metric system.

Decycling creates monetary value in two ways; from the carbon credits created when the trucks come off the road and from cost savings resulting when pickup schedules are reduced. As of early 2023, carbon credit systems are not fully developed, nor is there formal oversight or regulation to ensure credits are indeed credits. "Here's a new money-making model for you. Plant a small forest in your backyard. Call it 'afforestation' and 'carbon sequestration.' Calculate how many tons of carbon dioxide will be locked away in your forest over its lifetime. Then sell those carbon credits to companies and private entities who are still busy pumping CO_2 into the air. Congratulations, you've just marketed carbon offsets! It's not quite that easy, of course, but in the race to reduce their carbon footprint, companies are realizing that the carbon offset market is largely unregulated."[43]

Verifiable carbon credit creation centres on reducing the number of collection trips while tapping into existing infrastructure to move the bailed cardboard from the retailer back to central distributing. The number of emissions saved, illustrated in the following example, will be slightly higher or lower depending on the type of fuel used in the collection trucks. Gas has fewer emissions than diesel while natural gas has fewer emissions than gasoline.

Method one. Multiply the hours per month each truck is on the road by the number of trucks removed from the road to arrive at the total number of hours trucks are no longer operating. The total number of hours per month is then multiplied by emissions per hour to arrive at the total emission reduction per month. To get the yearly amount, multiply by 12.

Method two follows a similar path but uses distance no longer travelled multiplied by emissions per kilometre or mile. Exact savings calculations for each city or municipality will be based on the best available input data and could vary from the above methods.

REAL WORLD EXAMPLE

The following example is modelled on the city of Nanaimo, BC, population of roughly 100,000 in 2022. I selected this city because it was large enough to obtain meaningful measurements, the population size is easy to work with for scaling up to larger populations and Nanaimo is on record as being supportive of the tenets of a circular economy. The calculations on total emissions stopped were based in part on scheduling information and other data obtained from the official website for the city of Nanaimo.[44]

Trucks collect material from recycling bins at each home every second week. Garbage is also collected every other week alternating with recycling while food waste is collected weekly. Once Decycle phase one for cardboard is implemented, the recycling schedule gets reduced from every two weeks to once every four weeks which cuts the number of recycling truck trips from 24 to 12 per year. (Every other week year-round should equate to 26 pickups; however, two pickups per year do not occur due to the statutory holiday rotation.)

As noted, the food waste pickup schedule would also be reduced with no pickup in week two only. The new schedule becomes garbage and food waste week one, no trucks in week two, garbage and food waste in week three, with recycling and food waste in week four. The reduced schedule results in a 25% reduction in both truck traffic and carbon emissions. The corresponding reduction in city expenses is the total of all the savings, including reduced maintenance, labour, and fuel.

The 25 percent reduction in emissions can be calculated as a dollar amount by taking total emissions stopped in tonnes multiplied by the price of a tonne of carbon. These carbon credits can either be sold or applied as an offset against other city-based emissions in the pursuit of carbon neutrality for the city. The carbon credit price seems destined to increase in the future, governments have stated their intention to do so. The Canadian Federal Government has a schedule in place and on April 1, 2023, raised it by $15 from $50 to $65 per metric tonne. It was not a coincidence that gas at the pump went up by three cents a litre that day. (The rationale put forth by the Federal Government is by making it more costly to operate gas powered vehicles, consumers will be incentivized to switch to other transportation options like cycling, rapid transit and electric vehicles.)

In an extensive Canadian study published by arcplus. org, the authors calculated that each garbage and recycling truck operating year-round emitted **120 metric tonnes**

of CO$_2$.[45] According to the Recycling and Public Works Administration, the City of Nanaimo runs ten trucks a day, five days per week to pick up recycling, garbage and food waste bins, therefore, it can be estimated that total annual emissions for the 10 trucks (10 trucks x 120 metric tonnes) is 1,200 tonnes of carbon a year. After a Decycling program is in place, truck activity falls by 25%, resulting in yearly GHG savings of **300 metric tonnes.**

(10 trucks x 120 metric tonnes = 1200/4 = 300)

With the current carbon price in the province of British Columbia now at $65 per metric tonne, August 2023, total yearly carbon credits come to $65/tonne x 300 tonnes per year = **$19,500 per year or $195,000 per decade.** The decade calculation assumes the carbon price remains unchanged; however, it is slated to rise in Canada by $15 per metric tonne each year from 2023 to 2030 so the decade number is an underestimate if the increases are enacted as scheduled. Carbon credits created should be counted yearly since they are stopped not only in year one but in all subsequent years. Carbon capture systems operate continuously to pull carbon emissions from the atmosphere which get counted in that specific year.

City budget reductions could reach hundreds of thousands per year with larger cities climbing into the millions. No attempt was made to determine the cost savings as I did not request internal budget line items. The emissions stopped from Decycling become much larger when extrapolating out in a mass adoption scenario.

THE CASE FOR MASS ADOPTION

The assumption is that the above example will, for the most part, be applicable to cities and towns in North America, and Europe. In 2020, Canada's population was estimated at thirty-eight million which is 380 times the population size of Nanaimo.

The carbon emissions savings for Canada can be determined by scaling using (380 x 300 metric tonnes per year) to arrive at **114,000 tonnes of CO_2 emissions per year stopped.** Using a carbon price of $65/tonne equals **$7.4 million per year**.

Over a decade, the carbon emissions stopped comes to just over **1.1 million tonnes.** Using a carbon price of $65/tonne, the value of carbon credits comes out to **$71.5 million** per decade for Canada. Additional cost savings from the reduction in city budgets are not included but would add tens or possibly hundreds of millions more.

Once Decycling is implemented globally, the carbon emissions stopped becomes considerably larger given the number of cities and towns on the planet. For now, the focus is on the most developed western economies, Europe with 447 million people and the US with 332 million. When stage one Decycling is adopted in the US and EU, the numbers are:

*20.5 (population factor) x 114,000 tonnes/year CO_2 (Canada) = **2,337,000 tonnes of carbon per year**, no longer going into the atmosphere for Europe and the United States combined.

*(The scaling multiplication factor of 20.5 was derived from the population ratio of the US and Europe which adds

up to 779 million divided by the population of Canada, 38 million.) When the Canadian carbon emissions are added, total carbon emissions stopped comes to **2.45 million tonnes per year,** for Europe and North America combined and is a significant reduction that continues in perpetuity.

Carbon capture projects receive credits yearly for emissions captured by a carbon emitter; for stopped emissions to be correctly accounted for in the same way, the ensuing years from two and on should be counted accordingly.

ELECTRIC TRUCKS

Going all-electric for waste collection trucks is an ideal long-term goal; although these trucks exist today, they are not yet widely used. Electric collection trucks are extremely heavy due to the size of the battery pack required to power them. At least one waste management company is running electric trucks but at a weight of ten tons only have a range of 90 kilometres. "…electric garbage trucks are still few and far between."[46] These trucks can manage 600 pickups before requiring recharging. In the event that a truck completely discharges in the field, an on-call charging vehicle would be required to attend or they could be equipped with a back up gas generator to provide a charge sufficient for route completion.

Other heavy trucks are also being manufactured as all electric but doing so doesn't always make sense. Electric fire trucks have been delivered to a few departments in Canada but the range is low so they come equipped with a back up power system, a generator run on diesel. Having a building burn to the ground while the truck is sidelined a short distance away after depleting the battery is much more dire than not completing the garbage collection route on time.

There is also the fiscal reality of the higher cost of purchase, a challenge for any level of government disposing of taxpayers' funds. It makes sense to implement a Decycle programme, then sell the carbon credits created and earmark those funds along with city budget savings towards the purchase of an all electric fleet over time.

FALL OUT

Once Decycling is established, being able to discard excess material at the point of purchase may seem strange, most new programmes do. Some consumers may grumble about it slowing down shopping, but once people Decycle a few times, they will quickly become accustomed and will notice time savings as waste material at home is reduced. The adjustment may be similar to the introduction of beverage container deposits many years ago. The rationale then was to stop roadside pollution as refuse was tossed out of moving vehicles with little regard. At the time of introduction, some were agitated as they viewed the deposit as a tax, however, retailers and consumers quickly adjusted. The reality is that social acceptability shifts over time. For example, I can't recall the last time I saw a piece of trash being thrown out of a moving vehicle, most everyone today understands that doing so is simply not acceptable.

As consumers start to Decycle, they will feel more positive knowing they are participating in something practical that will reduce pollution and emissions. Once home collection frequency is reduced, less time will be needed for routine household tasks like gathering up all the recycling and getting the bin out to the curb on collection day. It's not that dragging the bins to the curb takes that long; it's just one more thing on the don't forget to-do list and the last-minute realization that the bins have not been taken out. There is no reason that an efficient society can't achieve once-a-month pickup where garbage, food waste, and recycling are all retrieved on the same trip. What I have outlined here is just the first step toward that, besides, overcollection may already be happening. In my neighborhood, many homes only have one or two occupants, and I have noticed that some residences don't always have bins out on collection day. It is likely that these households would not experience any impact from a reduction in pickup schedules even if they chose not to Decycle.

City and local governments should support Decycling since it will reduce waste management budgets. There are a variety of options for redirecting these savings, including green space enhancements, initiatives for the homeless or towards the purchase of an electric truck fleet. The monetary value of the carbon credits created are in addition and could be used in the same way or differently. The point is that all levels of government can facilitate the implementation of Decycling, including offering retailers incentives to participate. Federal, Provincial, and State governments have already committed to carbon neutrality, so supporting initiatives consistent with those goals seems obvious.

Initially, some retailers may view Decycling with scepticism and choose to narrowly focus on the increase in labour expense rather than the positive environmental benefits. The impact on retailers will be marginal, amounting to a few extra hours of labour to collect and bale the Decycled cardboard plus a nominal one time set up cost to create a Decycle point.

Lack of action thus far to right size and reduce packaging by the major food companies is probably due to bottom line concerns already covered, however, companies that actively engage in green initiatives are being held in higher regard by consumers and investors. Major "green" trends are in motion so reducing excess packaging is inevitable. It may be that in the future, some retail grocery stores will have a much wider array of self-serve options. For department stores, hardware stores, and others, Decycling volumes will be lower as fewer goods tend to be over-packaged while some items have little or no packaging. A few retailers may retain return policies that require the original packaging. If that is the case or if there is any doubt, the box goes home. Some retailers are already at zero excess packaging such as book sellers provided they only sell books.

Product suppliers and distributors pay fees to offset the waste created when their products are sold, so it is in their own best interest to take the initiative and reduce excess packaging. Key information currently displayed on the outside could be printed on the inner packaging. Having the internal packaging eventually become the outer packaging is attainable with many products.

To ensure home recycling bins do not fill up before four weeks, other excess cardboard should also be diverted including boxes from online shopping as these companies are in essence getting a free pass. There is no legitimate reason the online ordering companies can't pick up the cardboard boxes previously left behind or even better, stop over packaging delivered items in the first place. Pizza companies are decades behind on innovation and the implementation of new ways to get their food to the consumer without using the standard cardboard pizza box.

As was noted in the introduction, it's time to become more efficient as a society by taking a sober second look at what we have become accustomed to doing over the years. Chauffeuring our waste material all over town is a rather ridiculous activity when one pauses to look at it for what it is. Low cost solutions are available, Decycling savings are quantifiable, when those carbon-emitting trucks come off the road, the emissions are stopped.

AFTERWORD

The issues plaguing the earth right now go well beyond what has been covered in this short book. Hopefully, this contributes to a shift in mindset as many people continue to myopically champion one specific avenue with an over blown sense of confidence. The way forward hinges on individuals embracing the fact that it's up to them and the choices they make every day.

Pretending everything is fine as is or leaving it up to others for solutions is in my view, the crux of the problem and is how the earth got into this stressed state. In the current environment of rapid, divisive over the top communications, the fact that we are all on the same side seems to have been forgotten. The path forward involves working together and abandoning the my way is the only way mentality. The final verdict is out and will not be known for some time. If we can be honest and accept that nothing is 100 percent green, real conversations can take place. It is about trade-offs and compromises and taking steps in the right direction when and wherever feasible like using natural gas instead of wood or coal as opposed to the all or nothing approach.

Life is nothing more than a finite unknowable amount of time, as such, time really is the scarcest resource so the question becomes is it worth trading all those hours working to upgrade a one year old perfectly functioning cell phone? Undoubtedly it may be uncomfortable for some to look within at what is actually driving their buying decisions. As noted earlier, indiscriminate consumption is often a cover for unresolved personal issues.

Actions to take now include leaving cardboard at the retail store either at point of purchase or in a cart outside the store provided you do not leave a mess behind. Start sending the cardboard from online deliveries back with the delivery driver. Hand them over if at home when they arrive or place the previously left boxes outside the front door with a note instructing the driver to take back their boxes. It is their property, you did not order it, you only ordered the goods.

On Instagram follow Decyclegreen for interesting content and roll out progress updates. For more information or inquiries on consulting for your retail enterprise or for city managers email info@decycle.green.

ACKNOWLEDGEMENTS

My generation zed children have been key contributors and supporters. The first time I Decycled was as a direct result of becoming a father. The first few years I Decycled, they gave me a tough time about it but as they got older and understood it they became increasingly more supportive. Both deserve and get formal credit. My oldest son Daryl was the editor and content critic while my youngest son Andrew created the logo and set up some of the online tech related initiatives. I also want to thank all the other young people that were enthusiastically encouraging about Decycling. Thanks also to Jerry, who provided feedback early on. A special thanks to Taaj, the head of waste management for the City of Nanaimo, for taking time out of his busy schedule to read it and provide invaluable feedback.

REFERENCES

1 Helen Briggs, "Climate Change: The Environmental disaster we've almost fixed" bbc.com, November 7, 2022, bbc.com/news/science-environment-58874831

2 Helen Briggs, "Climate Change: The Environmental disaster we've almost fixed" bbc.com November 7, 2022, bbc.com/news/science-environment-58874831

3 Manfred Ledermuller "Walter Kreisel tackles the global challenge of water shortage" do-it.world/phantor/

4 Derya Ozdemir, "Transparent solar panels could replace windows in the future. Here's how" Interesting Engineering, April 27, 2022, interestingengineering.com/innovation/transparent-solar-panels-replace-windows-in-the-future-heres-how

5 Tiffany Crawford, "Lytton could be the first B.C. municipality to be powered by solar sidewalks" Vancouver Sun, June 24, 2022, vancouversun.com/news/local-news/lytton-could-be-first-b-c-municipality-to-be-powered-by-solar-sidewalks

6 Helene Ahlborg & Martin Sjosted, "Small-scale hydropower in Africa: Socio-technical designs for renewable energy in Tanzanian Villages" Elsevier – Energy Research & Social Science Volume 5, January 2015, Pages 20-33 sciencedirect.com/science/article/pii/S2214629614001510

7   Amanda Little, "Yes, we can make it rain. But it won't solve drought." Bloomberg Online, September 10, 2022, bloomberg.com/opinion/2022,cles/2022-09-10/climate-change-won-t-be-solved-with-cloud-seeding?leadSource=uverify%20wall

8   Staff writers, "Global Climate Change, Vital Signs of the Planet", Nasa website, updated November 2022, climate.nasa.gov/vital-signs/sea-level/

9   Rhett Herman, "How fast is the earth moving?" Scientific American, October 16, 1998, scientificamerican.com/article/how-fast-is-the-earth-mov/

10 Staff writers, "Environment: The Worlds Climate: Unpredictable" Time Magazine, August 9, 1976, content.time.com/time/subscriber/article/0,33009,914494,00.html

11 Michelle Butterfield, "Part of the sun breaks free and forms a strange vortex, baffling scientists" Global News, February 9, 202, globalnews.ca/news/9474534/solar-polar-vortex-sun-flare/

12 Quora contributor, "Why Looking At The Stars Is A Look Back In Time" Forbes online, February 7, 2018, forbes.com/sites/quora/2018/02/07/why-looking-at-the-stars-is-a-look-back-in-time/?sh=3f7d058214ec

13 Constantin W. Arnscheidt & Daniel H. Rothman, "Presence or absence of stabilizing Earth system feed-backs on different time scales", Science Advances, November 16, 2022, science.org/doi/10.1126/sciadv.adc9241

14 Susan Callery, "Global Warming vs Climate Change", Managing Editor, NASA Global Climate Change, March 28, 2023, climate.nasa.gov/global-warming-vs-climate-change/

15 Staff writers, "Closing The Loop On Pizza Box Waste", World Centric, October 26, 2018, https://www.worldcentric.com/journal/closing-the-loop-on-pizza-box-waste

16 Grace Hauck, "Where your recycling actually goes-and what you can do about it this Earth Day", phys.org, April 22, 2022, phys.org/news/2022-04-recycling-goesand-earth-day.html

17 Staff writers, "Plastic waste and pollution reduction", Government of Canada, March 1, 2023, canada.ca/en/environment-climate-change/services/managing-reducing-waste/reduce-plastic-waste.html

18 Angus Bennett, Producer & Editor, Kit Chellel, Original Reporting, "Tracking Devices Reveal Where Recycling Really Goes", YouTube Bloomberg Channel, March 29, 2022, youtube.com/watch?v=hmGrI_BVlnc

19 Antonella Artuso, "Beverage container recycling fee a 'pop can tax,' CTF says", Toronto Sun, February 10, 2023, torontosun.com/news/provincial/beverage-container-recycling-fee-a-pop-can-tax-ctf-says

20 Al Costa, "The Recycling Model Has Failed and Needs a Complete Revamp", tekntrash.ai, Best Startup London, October 3, 2022, beststartup.london/al-costa-the-recycling-model-has-failed-and-needs-a-complete-revamp/

21 Al Gore & Walter Isaacson, "Al Gore on Climate Change and Global Sustainability", Amanpour & Company transcript, November 17, 2022, pbs.org/wnet/amanpour-and-company/video/al-gore-on-climate-change-and-global-sustainability/

22 Allison Lampert, "Air Canada to buy 30 electric planes from Heart Aerospace", Reuters, September 15, 2022, reuters.com/business/aerospace-defense/air-canada-buy-30-electric-planes-heart-aerospace-2022-09-15

23 Mark P. Mills, "Mines, Minerals, and "Green" Energy: A Reality Check", Manhattan-Institute, July 9, 2020, manhattan-institute.org/mines-minerals-and-green-energy-reality-check#notes (Original attribution, Jeff Desjardins, visualcapitalist.com March 2, 2016)

24 Elias Huuhtanen, "Posts mislead about environmental impact of green energy", AFP Finland, AFP USA, March 31, 2022, factcheck.afp.com/doc.afp.com.327A429

25 Leigh Collins, "Shell has quietly closed down all its hydrogen filling stations in the UK", Hydrogeninsights, October 17, 2022, hydrogeninsight.com/transport/exclusive-shell-has-quietly-closed-down-all-its-hydrogen-filling-stations-in-the-uk/2-1-1335049

26 Staff writers, "Sustainable Energy: Recycling Renewables", Canadian Renewable Energy Association (CanREA) April 2021 fact sheet, renewablesassociation.ca/wp-content/uploads/2021/04/Recycling-Solar-Panels-English-Web.pdf

27 Dan Healing, "'The Gas City' pulls plug on money-losing $13-million solar power project", Canadian Press, Calgary Herald, May 16, 2019, calgaryherald.com/business/energy/the-gas-city-pulls-plug-on-money-losing-13-million-solar-power-project?_ga=2.101271938.244569910.1664907415-1684828601.1662825705&_gl=1*1ril95x*_ga*MTY4NDgyODYwMS4xNjYyODI1NzA1*_ga_72QH41ZTMR*MTY2NDkxNjMwNy4zLjEuMTY2NDkxNjQwMS4zMC4wLjA

28 Michelle Lewis, "Where do solar panels go when they die?", electrek.co, August 24, 2020, electrek.co/2020/08/24/where-do-solar-panels-go-when-they-die/

29 Peter Varga, "Diavik mine's wind-driven power saves $5 million a year in diesel costs", Nunatsiaq News, April 11, 2014, nunatsiaq.com/stories/article/65674diavik_mines_wind-driven_power_saves_5_million_in_diesel_costs/

30 Michael Shellenberger, "Why Wind Turbines Threaten Endangered Species With Extinction", Forbes.com June 26, 2019, www.forbes.com/sites/michaelshellenberger/2019/06/26/why-wind-turbines-threaten-endangered-species-with-extinction/?sh=217afaac64b4

31 Tom Leonard, "Graveyard of the green giants" Wind-watch. org, February 28, 2022, originally published by dailymail.com February 27, 2022, www.wind-watch.org/news/2022/02/28/ graveyard-of-the-green-giants/

32 Andrew Paul, "A wind turbine just smashed a global energy record-and it's recyclable", Popular Science, October 13, 2022, msn.com/en-ca/news/technology/a-wind-turbine-just-smashed-a-global-energy-record-and-it-s-recyclable/ar-AA 12VKEn?ocid=winp1taskbar&cvid=f0aad5bbe8db4e55a9a 0bb636333e659

33 David Klein, "Net zero versus real zero and the future of the planet", resilience.org, January 19, 2022, (originally published by System Change not Climate Change), resilience.org/stories/2022-01-19/net-zero-versus-real-zer o-and-the-future-of-the-planet/

34 Cheryl Santa Maria, "Study finds half of replanted trees don't survive", theweathernetwork.com, November 30, 2022, theweathernetwork.com/en/news/nature/outdoors/ study-finds-half-of-replanted-trees-dont-survive

35 Volker Quaschning & Bernhard Siegel, Specific Carbon Dioxide Emissions of Various Fuels", Volker-quaschning.de, November 2022, "volker-quaschning.de/datserv/CO2-spez/ index_e.php

36 Digital Writers, ""The way of the future": Run-of-the-river projects generate clean energy in B.C.", theweathernetwork. com, October 21, 2022, theweathernetwork.com/en/news/ climate/solutions/power-to-the-people-run-of-the-river-pro jects-generate-clean-energy-in-b-c

37 Staff writers, U.S. Energy Information Administration, "Nuclear Explained", November 7, 2022, eia.gov/ energyexplained/nuclear/nuclear-power-and-the-environment.php

38 Christina Nunez, "Micro-nuclear reactors: up to 20MW, portable, safer", energypost.eu, April 22, 2021, energypost.eu/micro-nuclear-reactors-up-to-20mw-portable-safer/

39 Staff writers, "Nuclear Power in the World Today", World Nuclear Association, updated March 2023, world-nuclear.org/information-library/current-and-future-generation/nuclear-power-in-the-world-today.aspx

40 David Dalton, "Sweden/New Government Announces Plans To Build Nuclear Power Plants", NUCNET, October 14, 2022, nucnet.org/news/new-government-announces-plans-to-build-nuclear-power-plants-10-5-2022

41 Caitlan Hester, "Is Recycling Cardboard Worth It? All The Facts (Easily Explained)", citizensustainable.com, December 8, 2021, citizensustainable.com/recycling-cardboard/

42 Jeffrey Currie, Tweet@SquawkCNBC, October 3, 2022, 6:07 a.m. twitter.com/SquawkCNBC/status/1576921977754902528

43 Staff writers, carboncredits.com, March 30, 2023 "Who Verifies Carbon Credits?" carboncredits.com/who-verifies-carbon-credits/

44 Staff writers, "Automated Waste Collection Services", City of Nanaimo, January 26, 2022, nanaimo.ca/city-services/garbage-recycling

45 Thuy T.T. Nguyen, "Assessment of fuel consumption and GHG emissions for the solid waste collection activities by using GPS data", University of New Brunswick, January 2008, www.acrplus.org/images/pdf/pdf_conference_waste_climate/document403.pdf

46 Jonathan M. Gitlin, "Seattle makes history with electric garbage truck", arstechnica.com, May 22, 2019, arstechnica.com/cars/2019/05/seattle-makes-history-with-electric-garbage-truck/

ABOUT THE AUTHOR

Robert Setter is an analyst, consultant and writer. He left UBC's faculty of science at age nineteen to work for a multinational retailer before completing a degree in economics a few years later. After a short stint as a federal investigator left him disillusioned with government inefficiencies, he embarked on a twenty-year career in education, eventually reaching director for a private college before pivoting to consulting and writing full time. This lead to several director positions with companies exploring the metals used in electric-vehicle batteries, which in turn led to a search for new ways to reduce emissions. The book, Decycle, introduces a new practical solution that will reduce carbon emissions and save money for taxpayers. He currently resides on the east coast of Vancouver Island in the province of British Columbia.